W9-AKY-889

G. Jakobi, A. Löhr

Detergents and Textile Washing

Distribution
VCH Verlagsgesellschaft, P.O. Box 12 60/12 80. D-6940 Weinheim
(Federal Republic of Germany)
Switzerland: VCH Verlags-AG, P.O. Box, CH-4020 Basel (Switzerland)
Great Britain and Ireland: VCH Publishers (UK) Ltd., 8 Wellington Court,
Wellington Street, Cambridge CB1 1HW (Great Britain)
USA and Canada: VCH Publishers, Suite 909, 220 East 23rd Street, New York
NY 10010-4606 (USA)

ISBN 3-527-26810-3 (VCH Verlagsgesellschaft) ISBN 0-89573-686-1 (VCH Publishers)

Detergents and Textile Washing

Principles and Practice

Günter Jakobi and Albrecht Löhr

in collaboration with
Milan J. Schwuger, Dieter Jung, Wilhelm K. Fischer, Peter Gerike, Klaus Künstler

(Henkel KGaA, Düsseldorf)

Dr. Günter Jakobi
Dr. Albrecht Löhr
Henkel KGaA
Development Detergents
D-4000 Düsseldorf

Translator: Dr. William E. Russey
Professor of Chemistry
Juniata College, Huntingdon (USA)
Editorial Director: Dr. Hans F. Ebel
Production Manager: Myriam Nothacker

Library of Congress Card Number 87-25299

British Library Cataloguing in Publication Data

Jakobi, Günter
Detergents and textile washing : principles and practice.
1. Detergents, Synthetic
I. Title II. Löhr, Albrecht
668′.14 TP992.5

ISBN 0-89573-686-1

Deutsche Bibliothek Cataloguing-in-Publication Data:
Jakobi Günter:
Detergents and textile washing : principles and practice /
Günter Jakobi and Albrecht Löhr. In collab. with Milan J.
Schwuger ... – Weinheim ; Basel (Switzerland) ; Cambridge ;
New York, NY : VCH, 1987
ISBN 3-527-26810-3 (Weinheim, Basel) Gb.
ISBN 0-89573-686-1 (Cambridge, New York) Gb.
NE: Löhr, Albrecht:

The cover illustration shows the structural model of crystalline zeolite 4A (courtesy of Henkel KGaA)
Composition, printing, and bookbinding: Graphischer Betrieb Konrad Triltsch, D-8700 Würzburg
Printed in the Federal Republic of Germany

Foreword

The present monograph is taken from a chapter entitled "Detergents" in *Ullmann's Encyclopedia of Industrial Chemistry* written by several employees of Henkel KGaA. The worldwide importance of this topic warrants the publication of a separate monograph which will reach a wide readership.

I am very pleased that this book has been prepared. It is intended not only for specialists in the field but for everybody interested in the textile washing process and its effects on the environment. Readers will gain a thorough understanding of the action of detergents and the many factors involved in the washing process, and it is thus my fervent hope that this book will provide an instructive contribution to the public discussion of the environmental effects of detergents.

In conclusion, I would like to express my gratitude to the colleagues of Henkel KGaA who were involved in the preparation of this monograph.

Prof. Dr. H. Verbeek
Corporate Vice President
Development Consumer Goods
Henkel KGaA

Preface

Detergents belong to the group of consumer products which are indispensable for the maintenance of cleanliness, health and hygiene. Their economic importance worldwide is considerable, although consumption varies markedly from country to country.

In previous years it was the consumers – mainly housewives – and those industries directly involved in the washing of textiles, namely, the textile industry, and washing machine and detergent manufacturers, who were interested in the topic "Washing of Textiles". However, recently there has been a growing interest in the subject among legislators and the general public. In an age of growing environmental concern, a change of attitude to the washing process has taken place in many countries. Today, the utilization of water, energy and chemicals has to be viewed not only from economic but also from ecological angles.

As a result, new raw materials, washing processes and washing technologies have been developed. Their common goal is to use the finite resources of our earth with care and so to prevent pollution of the environment as far as possible. These developments have led to the creation of many new terms. Moreover, as a result of coverage in the media, some terms which were previously only known to specialists are now familiar to wide sections of the general public.

The aim of the present book is to provide a comprehensive survey of all the parameters involved in the textile washing process, in particular the action of detergents. Further, the current state of the field worldwide is discussed and developments considered. The book is derived from a chapter entitled "Detergents" in *Ullmann's Encyclopedia of Industrial Chemistry* *. The physical and chemical principles involved in the washing process are described, as well as the composition, action and production of household and industrial detergents. For those readers interested in analytical aspects, a special chapter ("Analysis") presents a survey of modern methods used in detergent analysis. Throughout the book particular emphasis has been placed on ecological and toxicological aspects. A discussion of the economic importance of detergents and information on textile types and washing machines complete the book.

The reader interested in further details can refer to more than 600 references covering all aspects of the subject.

Because of the worldwide importance of detergents, we have not only considered their use in Europe but have also attempted to present a fair survey of the products and processes used outside of Europe, especially in the USA and Japan. Very large

---

* VCH Verlagsgesellschaft, Weinheim; Volume A8, 1987.

differences in washing processes and detergent formulations are observed, dependent in part on totally different washing habits.

In the first place, this book is intended for specialists in industry (e.g. producers of raw materials, manufacturers of detergents, washing machines and textiles), for universities (professors and students), schools, authorities, institutes, water and wastewater specialists, ecologists, toxicologists and dermatologists. Further, journalists, consumer organizations, environmental consultants and general readers will find it a valuable source of information on washing and its effects on the environment. The book will help them gain insight into this complicated process and the various factors involved.

We are especially indebted to the following colleagues who contributed chapters to this book:

Dr. Milan J. Schwuger, Chap. 2: Theory of the Washing Process; Dr. Dieter Jung, Chap. 6: Production of Powdered Detergents; Dr. Wilhelm K. Fischer, Dr. Peter Gerike, Chap. 10: Ecology; Dr. Klaus Künstler, Chap. 11: Toxicology.

Moreover, we would like to thank the translator, Prof. Dr. William E. Russey, and Dr. Donald L. Thompson for his careful copy-editing of the manuscript. We would like to offer our special thanks to Henkel KGaA for their support, and to the editorial team of Ullmann's, in particular Dr. Wolfgang Gerhartz who suggested we prepare this monograph. Finally, we thank Dr. Hans F. Ebel of VCH for the many helpful suggestions he made and for his help in the planning and preparation of the book.

This monograph is devoted to a subject of international importance, with the purpose of presenting the interested reader with a clear picture of the current state of knowledge in textile washing.

Düsseldorf
July 1987

Günter Jakobi
Albrecht Löhr

Contents

1 Historical Review

The symbol used by the ancient Egyptians to represent a launderer was a pair of legs immersed in water. This choice was logical, because at that time the standard way to launder clothes was to tread on them. The "fullones" of ancient Rome earned their bread, too, by washing clothes with their feet. The washing process then was very simple: laundry of every kind was subjected to purely mechanical treatment consisting of beating, treading, rubbing, and similar procedures. It has long been known, however, that the washing power of water can be increased in various ways. Rainwater, for example, was found to be more suitable for washing than normal well water. Hot water also was found to have more washing power than cold, and certain additives seemed to improve any water's effectiveness.

Even the Egyptians used soda as a wash additive. This was later supplemented with sodium silicate to make the water softer. These two substances formed the basis for the first commercial detergent brand to appear on the German market, Henkel's "Bleichsoda", introduced in 1878. Its water-softening effect resulted from the precipitation of calcium and magnesium ions, and it simultaneously eliminated iron salts, which had a tendency to turn laundry yellow. Used along with soap, which had also been known since antiquity, this product prevented the formation of inactive substances known as "lime soaps", and the laundry no longer suffered from a buildup of insoluble soap residues.

Soap is the oldest of the surfactants. It was known to the Sumerians by ca. 2500 B.C., although the Gauls were long credited with its discovery. For more than 3000 years, soap was regarded strictly as a cosmetic — in particular a hair pomade — and as a remedy. Only in the last 1000 years has it come to be used as a general purpose washing and laundering agent. Soap remained a luxury until practical means were discovered for producing the soda required for saponification of fats. With the beginning of the 20th century and the introduction of the first self-acting detergent (Persil, 1907), soap took its place as one ingredient in multicomponent systems for the routine washing of textiles. In these, soap was combined with so-called builders, usually sodium carbonate, sodium silicate, and sodium perborate. The new washing agents were capable of sparing people the weather-dependent drudgery of bleaching their clothes on the lawn, and the enhanced washing power of these new agents substantially reduced the work entailed in doing laundry by hand.

The next important development was the transition brought about by technology from the highly labor-intensive manual way of doing laundry to machine washing. This change in turn led to a need for appropriate changes in the formulation of washing agents. Soap, notorious for its sensitivity to water hardness, was gradually replaced by synthetic surfactants [1], [2] with their more favorable characteristics.

The first practical substitutes for soap were fatty alcohol sulfates, discovered in Germany by BERTSCH and coworkers in 1928 [3], [4]. The availability of synthetic alkyl sulfates based on natural fats made possible the first neutral detergent for fine washables: Fewa, introduced in Germany in 1932. This was followed in 1933 on the United States market by Dreft, a similarly conceived product. Fatty alcohol sulfates and their derivatives (alkyl ether sulfates, obtained by reacting fatty alcohols with ethylene oxide prior to sulfation) retain their importance currently in many applications, particularly in heavy-duty detergents, specialty detergents, dishwashing agents, cosmetics, and toiletries. The general acceptance worldwide of synthetic surfactants, with their reduced sensitivity to water hardness relative to soap, is a development of the years since World War II. Procter & Gamble introduced the synthetic detergent Tide into the United States in 1946. By the 1950s the widespread availability of tetrapropylenebenzenesulfonate (TPS), a product of the petrochemical industry, had largely forced soap off the cleansing agent market in the industrialized nations. Its only remaining role was that of a foam regulator. The favorable economics associated with TPS, along with its desirable properties, resulted by 1959 in this branched-chain synthetic surfactant's capture of ca. 65 % of the total synthetic surfactant demand in the Western world.

However, a new criterion soon appeared for surfactants, a criterion that had long been ignored: the biodegradability of the products. As a result of several unusually warm summers in Germany, especially that of 1959, water flow in many streams and rivers was severely restricted. Great masses of foam began to build up in the vicinity of dams and other obstructions. These were caused by insufficient biodegradation of both TPS and nonylphenol ethoxylates, a second class of detergents that had recently been introduced. The discovery that many surfactants could emerge unchanged even from a modern sewage treatment plant and, thus, enter the surface water led in 1961 to adoption of the first German Detergent Law [5], whose provisions took effect in 1964 [6]. Manufacturers were subsequently enjoined from marketing any detergents or cleansing agents whose biodegradability fell below 80 % in a test devised by the Detergent Commission. Initially only anionic surfactants were affected, but these were later joined by nonionic surfactants. The German precedent was soon followed by the enactment of similar legislation in such countries as France, Italy, and Japan. In the United Kingdom and the United States, the transition to biodegradable surfactants occurred as a result of voluntary agreements between industry and government.

Branched alkylbenzenesulfonates and nonylphenol ethoxylates have been largely replaced by the much more rapidly and effectively degraded linear alkylbenzenesulfonates and fatty alcohol ethoxylates.

Another important step in the development of detergents was replacement of builders like sodium carbonate by complexing agents. The first complexing agents that were used were of the sodium diphosphate type, but these were replaced after World War II by the more effective sodium triphosphate. Recently, inorganic ion exchangers such as zeolite 4 A have also been introduced as builders. Table 1 summarizes the performance according to several criteria of detergents containing various builders and surfactants [7].

Table 1. Properties of detergents based on different surfactants and builders

Detergent	Single wash cycle performance	Soil antiredeposition capability	Deposition on fabrics and washing machines	Yellowing and bad odor
Soap–sodium carbonate–sodium silicate	poor	poor	very heavy	heavy
Synthetic nonionic surfactants–sodium carbonate–sodium silicate	fair to good	fair	heavy	none
Synthetic surfactants–sodium diphosphate	good	fair to good	heavy	none
Synthetic surfactants–sodium triphosphate	good	good	weak	none
Synthetic surfactants–sodium triphosphate–zeolite 4A	good	good	weak	none
Synthetic surfactants–zeolite 4A–polycarboxylates	good	good	weak	none

The years after World War II also saw the introduction of other ingredients for improving detergency performance, and their presence is currently accepted as belonging to the state of the art for modern cleansing agents. Chief among these are the following:

soil antiredeposition agents
enzymes
fluorescent whitening agents
foam regulators
bleaching activators

Table 2 shows the historical development of surfactants and detergents with the concurrent developments in textile fibers and washing machines. Comparison of the changes clarifies how closely the various participants in the washing process are tied to one another. The laundry occupies the center of the stage, and it has consistently kept pace both with detergents and with the available washing processes and machines, at least to the extent that the newly developed textiles have been found to have value and market appeal. Thus, close and sympathetic cooperation among the various manufacturers that are involved has become essential for the development of optimal washing processes.

Table 2. Development of detergent ingredients, detergents, textile fibers, and washing equipment from 1876 to 1984

Year	Detergent ingredients	Detergents	Textile fibers	Washing equipment
1876	sodium silicate soap starch		cotton linen wool	boiler
1878	sodium carbonate sodium silicate	prewashing product and laundry softener (Henko, Henkel, Germany)		
1890			cupro	
1907	soap sodium carbonate sodium perborate sodium silicate	heavy-duty detergent (Persil, Henkel, Germany)	rayon acetate silk	wooden vat machine
1913	proteases (pancreatic enzymes)	prewashing product (Burnus, Röhm & Haas, Germany)		
1920			viscose staple fiber	metal tub agitator washing machine
1932/33	synthetic surfactants (fatty alcohol sulfates)	specialty detergent (Fewa, Henkel, Germany; Dreft, P & G, USA)		
1933	sodium diphosphate magnesium silicate	heavy-duty detergent (Persil, Henkel, Germany)		
1940	alkylsulfonates (Mersolat) antiredeposition agents (CMC[a])	heavy-duty detergent (Henkel, Unilever, Germany)		automatic agitator washing machine (Blackstone, USA)
1946	fatty alcohol sulfates alkylbenzenesulfonates sodium triphosphate	heavy-duty detergent (Tide, P & G, USA)	polyamide	automatic drum-type machine (Bendix, USA)
1948	nonionic surfactants	heavy-duty detergent (All, Monsanto, USA)		
1949	fluorescent whitening agents (optical brighteners)	rinsing additive (Sil, Henkel, FRG)		

Table 2. (continued)

Year	Detergent ingredients	Detergents	Textile fibers	Washing equipment
1950	fragrances	heavy-duty detergent (Dial, Armour Dial, USA)		
	cationic surfactants	fabric softener (CPC International, USA)		
1954	anionic–nonionic combinations		polyacrylonitrile	semiautomatic drum-type machine (washing automatically, rinsing by hand, spinning separately, Europe)
	suds-controlling agents (soap)	heavy-duty detergent (Dash, P & G, USA)		
1957			polyester resin-finished cotton	automatic drum-type machine (washing and rinsing automatically, spinning separately, Europe)
1962	suds-controlling agents (behenate soap)	heavy-duty detergent (Dash, P & G, FRG)	polyester/ cotton blend	fully automatic drum-type machine (washing, rinsing, and spinning automatically, Europe)
1963	proteases (microbe-based enzymes)	prewashing product and laundry softener (Biotex, Kortmann & Schulte, Netherlands)		
1965			polyurethane wool with reduced felting	wash dryer (washing, rinsing, spinning, and tumble drying automatically, Europe)
1966			nonwoven	
1970	fatty acid amine condensation product	detergent for delicates with fabric softening effect (Perwoll, Henkel, FRG)	resin-finished linen	

Table 2. (continued)

Year	Detergent ingredients	Detergents	Textile fibers	Washing equipment
1972	bleach activators (TAGU)[b]	heavy-duty detergent (Cid, Henkel, FRG)		
	NTA[c]	heavy-duty detergent (various brands, Canada)		
1975	sodium citrate	heavy-duty liquid detergent without phosphate (Wisk, Unilever, USA)		
1976	zeolite 4 A	heavy-duty detergent (prodixan, Henkel, FRG; Tide, P & G, USA)		
	suds-controlling agents (silicon oils)	heavy-duty detergent (Mustang, Henkel, FRG) heavy-duty liquid detergent without builders (Era, P & G, USA)		
1977	layered silicates–cationic surfactants	dual-function detergent with fabric-softening effect (Bold 3, P & G, USA)		
1978			Dunova (modified polyacrylonitrile fiber)	microcomputer operation, electronic sensing (Europe)
1982	zeolite 4 A–NTA	heavy-duty detergent, phosphate-free (Dixan, Henkel, Switzerland)		
1984	poly(acrylic acid), poly(acrylic acid-*co*-maleic acid)	heavy-duty detergents		

[a] CMC = carboxymethyl cellulose. [b] TAGU = tetraacetylglycoluril. [c] NTA = nitrilotriacetic acid.

2 Theory of the Washing Process

Washing and cleansing in aqueous wash liquor is a complex process involving the cooperative interaction of numerous physical and chemical influences. In the broadest sense, washing can be defined as both the removal by water or aqueous surfactant solution of poorly soluble residues and the dissolution of water-soluble impurities.

A fundamental distinction exists between the primary step, in which soil is removed from a substrate, and secondary stabilization in the wash liquor of dispersed or molecularly dissolved soil. The goal of the latter is to prevent redeposition onto the fibers of soil that has already been removed. The terms *single* and *multiple wash cycle performance* are used respectively in conjunction with the two phenomena.

The following components constitute a partnership in the overall washing process:

water
soil
textiles
washing equipment
detergent

Wash performance is highly sensitive to such factors as textile properties, soil type, water quality, washing technique (amount and kind of mechanical input, time, and temperature), and detergent composition. Not all these mutually interrelated factors are amenable to random variation; indeed, they are generally restricted within rather narrow limits. Of particular importance is the constitution of the detergent.

2.1 Influence of the Water

The most obvious role of water is to serve as a solvent, both for the detergent and for soluble salts within the soil. Water is also the transport medium for dispersed and colloidal soil components, however. The washing process begins with wetting and penetration of the soiled laundry. Water has a very high surface tension (72 mN/m), and wetting can only take place rapidly and effectively if this is drastically reduced by surfactants, which thus become key components of any detergent.

Water hardness also has a significant influence on the results of the washing process. Water hardness is defined in terms of the amount of calcium and magnesium

Table 3. Units currently used for expressing the hardness of water [8]

Name of unit	Definition	Symbol	Conversion factors						
			Ca^{2+}		CaO	$CaCO_3$			
			mmol/L	meq/L	°d	mg/kg*	°e	°a	°f
Millimole per liter	1 mmol of calcium(II) ions (Ca^{2+}) in 1 L of water	mmol/L	1	2.000	5.600	100	7.020	5.8500	10.00
Milliequivalent per liter	20.04 mg of calcium(II) ions (Ca^{2+}) in 1 L of water	meq/L	0.500	1	2.800	50	3.510	2.9250	5.00
German degree of hardness	10 mg of calcium oxide (CaO) in 1 L of water	°d	0.178	0.357	1	17.8	1.250	1.0440	1.78
Milligram per kilogram	1 mg of calcium carbonate ($CaCO_3$) in 1 L of water	mg/kg*	0.010	0.020	0.056	1	0.070	0.0585	0.10
English degree of hardness	1 grain of calcium carbonate ($CaCO_3$) in 1 gal (UK) of water	°e	0.142	0.285	0.798	14.3	1	0.8290	1.43
American degree of hardness	1 grain of calcium carbonate ($CaCO_3$) in 1 gal (US) of water	°a	0.171	0.342	0.958	17.1	1.200	1	1.71
French degree of hardness	1 mol (= 100 g) of calcium carbonate ($CaCO_3$) in 10 m^3 of water	°f	0.100	0.200	0.560	10.0	0.702	0.5850	1

* The unit "part per million" (ppm) is often used for mg/kg.

Table 4. Distribution of water hardness shown as percentage of homes affected by defined ranges of hardness [9]

Country	Range of hardness		
	0–90 ppm	90–270 ppm	> 270 ppm
Japan	92	8	0
United States	60*	35	5
Western Europe	9**	49**	42**
Austria	1.8	74.7	23.5
Belgium	3.4	22.6	74
France	5	50	45
Federal Republic of Germany	10.8	41.7	47.5
Great Britain	1	37	62
Italy	8.9	74.7	16.4
The Netherlands	5.1	76.1	18.8
Spain	33.2	24.1	42.7
Switzerland	2.8	79.7	17.5

* Including 10% with home water softening appliances.

** Calculated from the data for the countries indicated below with regard to their individual population figures.

salts present, measured in millimoles per liter (mmol/L). A calcium hardness of 1 mmol/L corresponds to 40.08 mg of calcium ions per liter of water. Additional hardness data are provided in Table 3 [8], along with other measures of hardness and relationships among them.

Water hardness can vary considerably from one country to another (Table 4). Soft water is relatively rare in most of Europe, whereas it is the rule in the United States and Japan. Water hardness can also be subject within a given country to wide temporary and permanent variations, as illustrated by data from the United States (Table 5).

Water of poor quality can severely impair the washing process and have detrimental effects on washing machines. The calcium and magnesium ions, which are responsible for water hardness, are prone to precipitate, either in the form of carbonates or as insoluble compounds derived from substances present in detergents. These precipitates can lead to residues in the laundry, but they can also build up as scale in the washing machine, thereby impeding the function of electrical heating coils and other machine components. A high calcium content in the water also hinders removal of pigment soil. The presence of trace amounts of iron, copper, or manganese ions in water can also be detrimental to washing. For example, these ions can catalyze decomposition of bleaching agents during the washing process. Complexing agents or ion exchangers are often found in detergents, and one of their functions is to bind multivalent alkaline-earth and heavy-metal ions through chelation or ion exchange.

Table 5. Water hardness distribution in the United States*

	Percent of population			
	Soft water (0–60 ppm)	Medium water (61–155 ppm)	Hard water (156–260 ppm)	Very hard water (> 260 ppm)
Alabama	45	55		
Arizona	3	5	84	8
Arkansas	80	17	3	
Connecticut	100			
California	16	57	17	10
Colorado	66	13	21	
Delaware	79	21		
District of Columbia		100		
Florida	13	61	24	2
Georgia	83	14	1	2
Idaho	20	56	21	3
Illinois	1	81	5	13
Indiana		35	37	28
Iowa		50	19	31
Kansas	3	61	22	14
Kentucky	4	96		
Louisiana	16	81	3	
Maine	100			
Maryland	100			
Massachusetts	96	3		1
Michigan	1	79	10	10
Minnesota	8	65	6	21
Mississippi	91	5	4	
Missouri	1	83	12	4
Montana	20	78	2	
Nebraska		49	36	15
Nevada	36	4	54	6
New Hampshire	93	7		
New Jersey	57	33	8	2
New Mexico	8	55	14	23
New York	67	17	2	14
North Carolina	97	3		
North Dakota		56	28	16
Ohio	1	81	11	7
Oklahoma	8	79	8	5
Oregon	98	2		
Pennsylvania	30	57	10	3
Rhode Island	75	25		
South Carolina	99	1		
South Dakota		25	27	48
Tennessee	42	57	1	
Texas	31	42	20	7

Table 5. (continued)

	Percent of population			
	Soft water (0–60 ppm)	Medium water (61–155 ppm)	Hard water (156–260 ppm)	Very hard water (>260 ppm)
Utah		18	76	6
Vermont	75	25		
Virginia	49	48	2	1
Washington	82	18		
West Virginia	45	40	9	6
Wisconsin	5	63	4	28
Wyoming	12	35	25	28
National average	40	40	13	7

* Water hardness data source: U.S. Geological Survey; population data source: U.S. Census, 1972.

2.2 Types of Soil

One way to categorize soils is by their origin, i.e.:

dust from the atmosphere
bodily excretion
impurity derived from domestic,
commercial, or industrial activity

From a detergency standpoint, however, it is more appropriate to regard the principal types of soil in other ways. Thus, one can distinguish the following:

Water-soluble materials:
inorganic salts
sugar
urea
perspiration

Pigments:
metal oxides
carbonates
silicates
humus
carbon black (soot)

Fats:
animal fat
vegetable fat
sebum

mineral oil
wax

Proteins from the following:
blood
egg
milk
skin residues

Carbohydrates:
starch

Bleachable dyes from the following:
fruit
vegetables
wine
coffee
tea

Soil removal during the washing process is enhanced by increases in mechanical input, wash time, and temperature. For any given washing technology, however, detergency performance is dependent on specific interactions among substrate surface, soil, and detergent components. In this context, it is important to consider not only the interactions of the wash components with one another, but also interactions among the various classes of soil. The most difficult soils to remove from fabrics are pigments, such as carbon black, inorganic oxides, carbonates, and silicates. Other problem soils include fats, waxes, higher hydrocarbons, denatured protein, and certain dyes. All of these tend to be present on fibers in the form of mixed soils.

The removal of soil from a surface either can be coupled with a chemical reaction or it can occur without chemical change. A redox process involving a bleach is an example of the former, in the course of which some oxidizable substance (e.g., a natural dye from tea, wine, or fruit juice) is cleaved. Enzyme-mediated decomposition of a strongly bound protein soil and its subsequent removal would be another example.

In many cases, however, the soil to be removed consists of substances that are not amenable to chemical treatment, and the only alternative is to dislodge the soil by interfacial processes. This requirement is reflected in the composition of modern detergents. Apart from bleaches, the primary components of modern detergents are surfactants, water-soluble complexing agents, and, more recently, water-insoluble ion exchangers.

2.3 Physical Soil Removal

Physical removal of soil from a support occurs as a result of nonspecific adsorption of surfactants on the various interfaces present [10] and through specific adsorption of chelating agents on certain polar soil components [7]. In addition, an indirect effect is caused by calcium ion exchange, whereby the freeing of calcium ions from soil deposits and fibers leads to a loosening of any remaining residue [11]. Compression by electrolytes of the electrical double layer at boundary surfaces is also significant [10]. All of these effects work together to dislodge oily and pigmented soils from textile substrates or solid surfaces. Interfacial properties, which are modified by the adsorption of detergent components onto the various interfaces present, include the following:

Air–water interface:
surface tension
foam generation
film elasticity
film viscosity

Liquid–liquid interface:
interfacial tension
interfacial viscosity
emulsification
electric charge
active ingredient penetration

Solid–liquid interface:
splitting pressure
suspension stability
electric charge

Solid–solid interface:
adhesion
flocculation
heterocoagulation
sedimentation

Interfaces in multicomponent systems:
wetting
rolling-up processes

Interfacial chemical properties undergo change as detergent compounds are adsorbed, and these changes are prerequisite to effective soil removal. The significance of a given change in respect to the overall process can vary, depending on the system involved. However, one can say generally that the greater the equilibrium adsorption of washing active substances, and the more favorable their adsorption kinetics, the better will be their detergency performance [10].

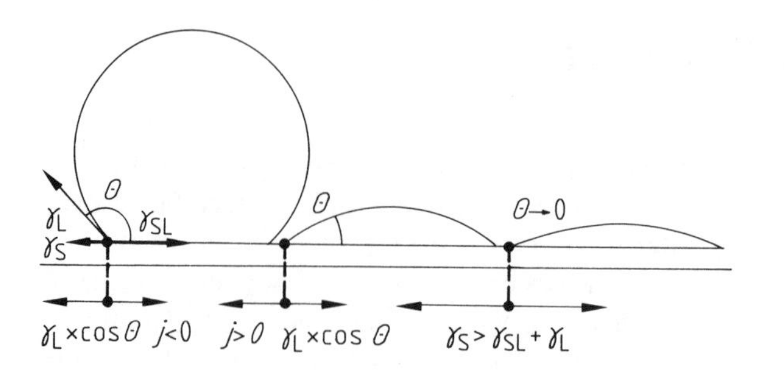

Figure 1. Schematic representation of wetting of a solid surface

2.3.1 Oily/Greasy Soil

Rolling-up Processes. Most oily and greasy soils are liquid at wash temperatures > 40 °C. Thermoanalytical investigations have shown that even fats that are solid at room temperature contain substantial amounts of liquid material [12]. These substances wet most textile substrates very effectively, and they have a tendency to spread over a surface, forming more or less closed covering layers. For this reason the following observations can be regarded as equally applicable to both liquid and solid fat residues.

In the first phase of washing, textile fibers and soil must be wetted as thoroughly as possible by the wash liquor [13]. The contact angle between a solid and a drop of a liquid applied to its surface can be taken as a measure of wetting. Figure 1 schematically depicts the way in which this angle decreases with decreasing surface tension γ_L.

To a first approximation, wetting can be described [14] by the Young equation [10]:

$$j = \gamma_S - \gamma_{SL} = \gamma_L \cos\theta \quad (1)$$

j = wetting tension (mN/m)
γ_S = interfacial tension substrate/air (mN/m)
γ_{SL} = interfacial tension substrate/liquid (mN/m)
γ_L = interfacial tension liquid/air (mN/m)
θ = contact angle in the wetting liquid

Total wetting of a solid is possible only if the liquid drop spreads spontaneously over the solid surface, e.g., when $\theta = 0$ and $\cos\theta = 1$. With a given solid surface possessing a low surface energy, for various liquids a linear relationship normally exists between $\cos\theta$ and surface tension [15]. The limiting value for $\cos\theta = 1$ is a constant of the solid and is referred to as its critical surface tension, γ_c. This means that only those liquids with surface tension equal to or less than the critical surface tension of a given solid spread spontaneously, thus causing thorough wetting. Table 6 contains a collection of critical surface tension data for various synthetic materials.

Table 6. Critical surface tensions of several typical plastics [16]

Polymer	γ_c at 20 °C, mN/m
Polytetrafluoroethylene	18
Polytrifluoroethylene	22
Poly(vinyl fluoride)	28
Polyethylene	31
Polystyrene	33
Poly(vinyl alcohol)	37
Poly(vinyl chloride)	39
Poly(ethylene terephthalate)	43
Poly(hexamethylene adipamide)	46

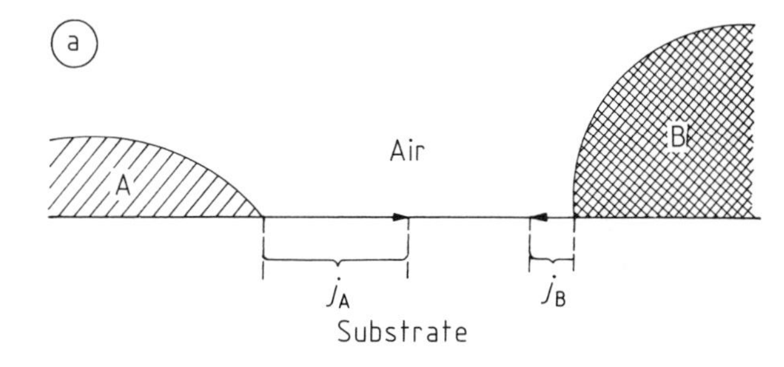

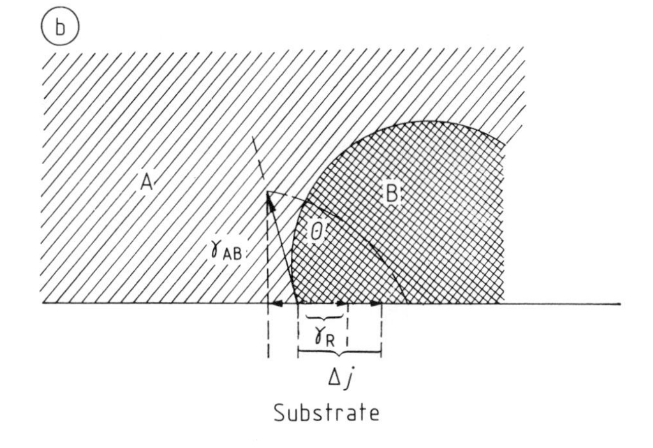

Figure 2. Two liquids on a solid surface
a) Separated; b) Overlapping
A) Wash liquor; B) Oily dirt

Polyamide, for example, which has a γ_c of ca. 46 mN/m, is relatively easy to wet with standard commercial surfactants, whereas polytetrafluoroethylene, which has a γ_c of ca. 18 mN/m, is wettable only with special fluoro surfactants. This relationship simplifies the choice of a proper surfactant for a given wetting problem.

Strictly speaking, the limiting case described above, that of total wetting (Eq. 1), is only applicable if γ_{SL} is so reduced through adsorption that it approaches 0. In typical washing and cleansing operations, the situation is much more complicated, since the solid surfaces that are present tend to be irregularly covered with oily or greasy soils. Thus, the wash liquor must compete with soil in wetting the surface. Figure 2 illustrates the problem schematically.

If two drops of different liquids (e.g., wash liquor A and an oily residue B) are placed next to each other on a solid surface S, two different wetting tensions j_A and j_B act on the surface. If the two liquids come into direct contact and form a common boundary surface, then the difference in their wetting tensions, Δj, the so-called oil displacement tension, is effective along the line of contact. In addition, a portion of the interfacial tension γ_{AB} remains effective along the substrate surface, but with reversed sign and a magnitude of $\gamma_{AB} \cos\theta$, where θ is the contact angle in the oily phase B. The total force that results is the contact tension, γ_R, given by Equation 2:

$$\gamma_R = \Delta j + \gamma_{AB} \cos\theta \qquad (2)$$

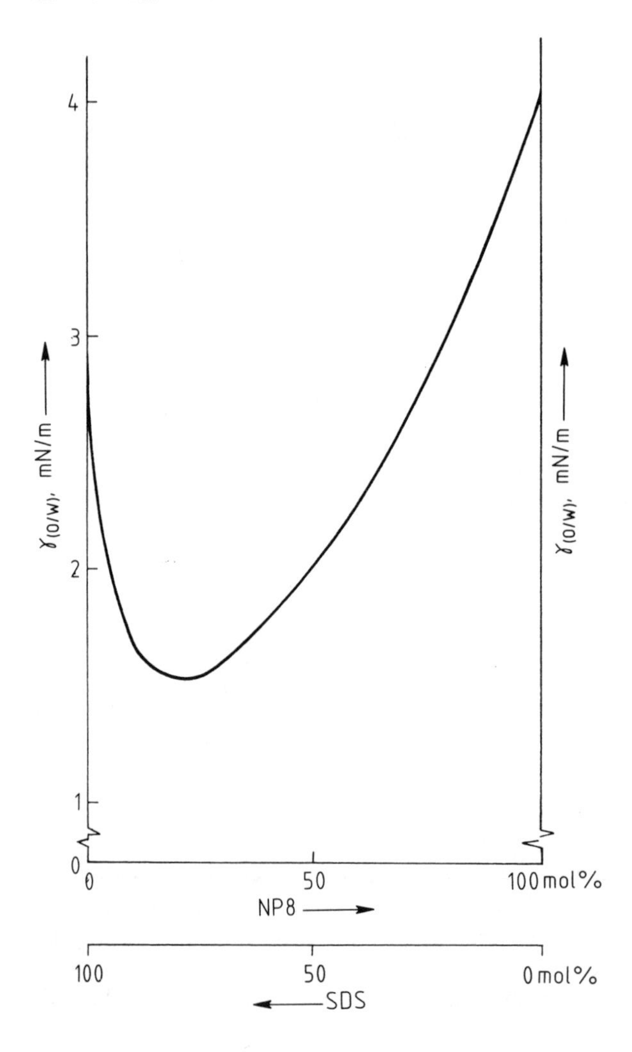

Figure 3. Influence of composition of a surfactant mixture on the interfacial tension of water–olive oil
Concentration: 1×10^{-3} mol/L; temperature: 30 °C; water hardness: 8 °d
SDS Sodium dodecyl sulfate; NP8 Nonylphenol octaethylene glycol ether

The contact tension acts in such a direction as to constrict the oil drop [17]. As a result of adsorption of surfactants out of phase A, the value of Δj is increased and that of γ_{AB} is decreased. The value of $\gamma_{AB} \cos \theta$ is negative for obtuse contact angles. From Equation 2, it can be seen that both quantities contribute to an increase in the contact tension and, thus, to better penetration of the oil drop. This complex process, in which a surface undergoes wetting first by oil and then by water, is known as "roll-up". For many cases of practical interest, roll-up is not a spontaneous phenomenon. More often, total constriction of an oil drop can be achieved only if mechanical energy is applied to the system [18]. The required energy, designated A_R (from the German "*R*estwasch*a*rbeit"), is directly proportional to the interfacial tension γ_{AB} and decreases with increasing surfactant concentration.

All the relationships presented thus far confirm the premise that interfacial tension is the primary force resisting the removal of liquid soils, and this force must be minimized if the washing process is to be effective. One way to reduce the interfacial tension is to create appropriate mixed adsorption layers comprised of surfactants of differing constitution [19]. For example, Figure 3 shows the interfacial tension of the system water–olive oil as a function of composition for a surfactant mixture containing the anionic surfactant sodium *n*-dodecyl sulfate with the nonionic surfactant nonylphenol octaethylene glycol ether, where the total surfactant concentration is kept constant [20].

Even small additions of one surfactant to another can lead to a significant reduction in the interfacial tension. A minimum value of interfacial tension is reached with a ratio of anionic surfactant to nonionic surfactant of ca. 4:1. The presence of a small

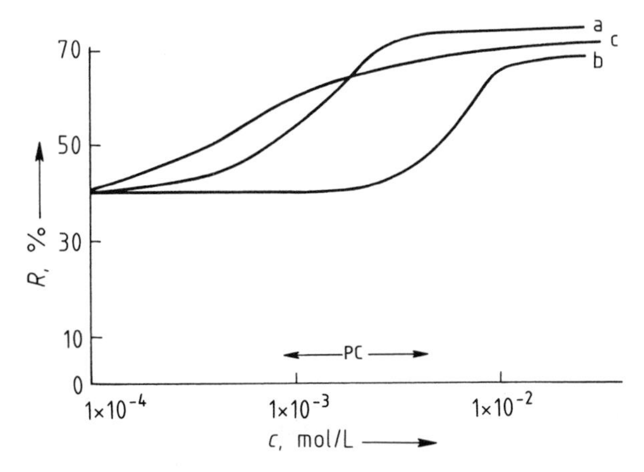

Figure 4. Wash performance expressed in terms of the remission *R* for mixtures of sodium *n*-dodecyl sulfate (SDS) and nonylphenol octaethylene glycol ether (NP8) [19]
a) NP8; b) SDS; c) Mixture of SDS/NP8 = 4:1 (plotted as NP8 concentration)
PC Concentration range corresponding to typical applications
Experimental conditions. Soil: mixed soils comprised of sebum, carbon black, kaolin, and iron oxide; water: deionized; temperature: 30 °C; bath ratio: 1:12; fabric: resin-finished cotton; apparatus: Launder-ometer

amount of nonionic surfactant in the surface layer reduces the mutual repulsion of negatively charged groups on the anionic surfactant, as a result of which adsorption increases. This is especially evident and important at low surfactant concentration, i.e., below the critical micelle concentration, as shown by wash studies (Fig. 4) using the previously described medium [20].

Findings such as these are independent of the nature of the fiber and apply to all soils containing hydrophobic pigments and oily material [21]. As a result, most commercial detergents contain mixtures of surfactants in carefully determined proportions.

Emulsification. In addition to roll-up, the emulsification of oily liquids and greases can also play a role in the washing process, provided certain specific preconditions are met. In particular, the system must possess a sufficiently high interfacial activity so that the interfacial tension falls below 10^{-2}–10^{-3} mN/m for soil to be dislodged by the largely substrate-nonspecific emulsification process [22]. However, these conditions are rarely observed in practice. As a result, emulsification is currently a major factor only in multiple wash cycle performance. In this case, the formation of largely stable emulsions inhibits the redeposition of previously removed liquid soils onto textile fibers.

Solubilization. Increasing surfactant concentration leads to a decrease in both surface tension and interfacial tension until the point is reached at which surfactant clusters begin to form. Above this concentration (the critical micelle concentration, c_M) changes in surface and interfacial activity are only minimal. Oil removal values also reach their upper limit at the critical micelle concentration of the surfactant

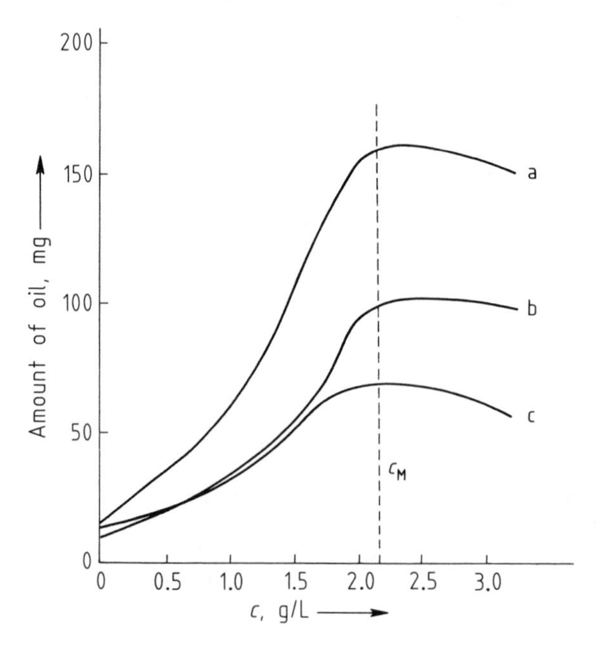

Figure 5. Removability of an olive oil–oleic acid mixture from wool, shown as a function of sodium *n*-dodecyl sulfate concentration for various degrees of oil coating (c_M critical micelle concentration) [23]
a) 4.6% Oil coat; b) 3.3% Oil coat; c) 2.2% Oil coat

employed [23]. This is illustrated in Figure 5 for the example of removing olive oil from wool.

From this behavior, one can conclude that effective washing is a result of the properties of individual surface-active ions, not of micelles. Just as with emulsification, however, micelles are able to solubilize water-insoluble materials and thereby prevent redeposition of previously removed soil in the later stages of the washing process. Values reported in the literature for critical micelle concentrations are based on pure solutions of surfactant. However, in an actual washing process, surfactants are invariably adsorbed onto a variety of surfaces, which means that the true surfactant concentration in the wash liquor is correspondingly diminished. The true surfactant concentration in solution is the only significant measure.

Formation of Mixed Phases. Penetration of individual detergent components (primarily surfactants) into an oily phase and the resulting formation of new anisotropic mixed phases can also cause a change in the water–oil interfacial tension. Development of liquid–crystalline mixed phases can be observed, for example, with olive oil–oleic acid–sodium *n*-dodecyl sulfate, and these lead to improved soil removal from textile substrates [23].

The dramatic effect of liquid–crystalline mixed phase formation on soil removal is illustrated in Figure 6 for the model system decanol–potassium octanoate. This combination forms such phases in the presence of added electrolytes [24].

Decanol is removed completely from the fibers, provided the wash experiment is carried out under conditions leading to liquid–crystalline layers at the boundary surfaces. When this is not the case, only minimal separation of the oily residue from the polyester fibers is observed, which is a result of a slow rolling-up process.

Specific Electrolyte Influences. In general, electrolytes are found to have only an indirect effect, and then only when anionic surfactant adsorption occurs at boundary surfaces. The addition of electrolytes leads to a compression of the electrical double layer at all boundary surfaces, which causes enhanced surface adsorption of surfactants. However, if the system is devoid of surfactants, a vast difference exists between

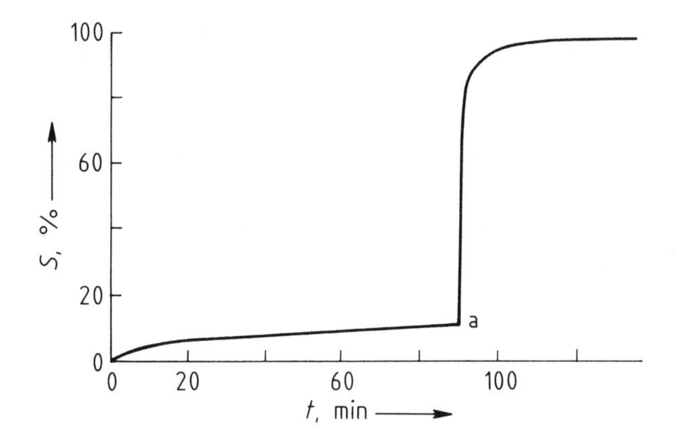

Figure 6. Washing effect S of potassium octanoate (2 wt%) on polyester fibers coated with decanol [24]
0.5 mol/L KCl added after 90 min
Prior to a: no formation of liquid–crystalline mixed phase; subsequent to a: formation of liquid–crystalline mixed phase

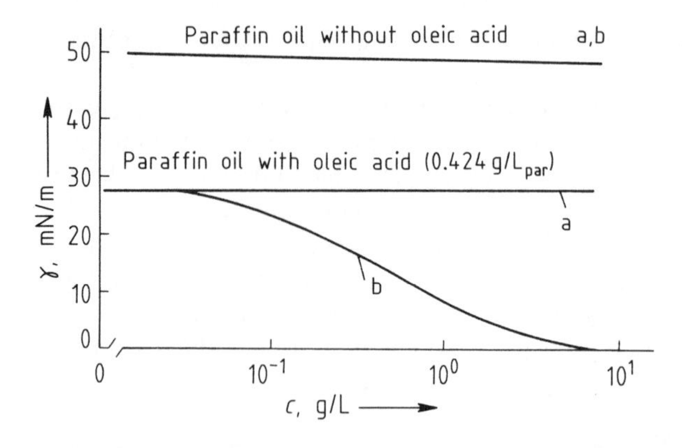

Figure 7. Influence of electrolytes on interfacial tension: paraffin oil–doubly distilled water at 25 °C
a) Na_2SO_4; b) $Na_5P_3O_{10}$

the properties of normal electrolytes, such as sodium chloride and sodium sulfate, and electrolytes prone to form complexes, including sodium triphosphate and sodium citrate. The differences in mechanism of action are apparent from Figure 7 [25].

In systems containing highly nonpolar oils, both sodium sulfate and sodium triphosphate show indifferent behavior, and they have no effect on the interfacial tension. When a small amount of oleic acid is added to the paraffin oil, however, the difference becomes clearly apparent. Although sodium sulfate has no effect on the interfacial tension, this tension is reduced dramatically by sodium triphosphate. Thus, electrolytes capable of forming complexes facilitate penetration of fatty acids through the interface, causing the liquid–liquid interfacial tension to be reduced. As a consequence, these materials are capable of activating a surfactant, an effect particularly significant for the removal of sebum, a substance rich in fatty acids. Table 7 shows that the effect is not simply soap formation in an alkaline medium. The interfacial tension observed with sodium and potassium hydroxide is significantly greater than that with sodium triphosphate, even though the pH in the latter case is lower.

The influence of a change in the oil–water interfacial tension on soil removal for oily liquid soils is demonstrated by model wash experiments carried out with the system sebum–polyester/cotton fabric (Fig. 8).

Table 7. Interfacial tension with sebum in surfactant-free alkali and sodium triphosphate solutions at 40 °C

Substance	Concentration, g/L	Interfacial tension, mN/m	pH	Water hardness, °d
NaOH	0.14	0.8	10	16
KOH	0.21	0.9	10	16
$Na_5P_3O_{10}$	2.0	0.09	9	16
$Na_5P_3O_{10}$	2.0	0.07	9	0

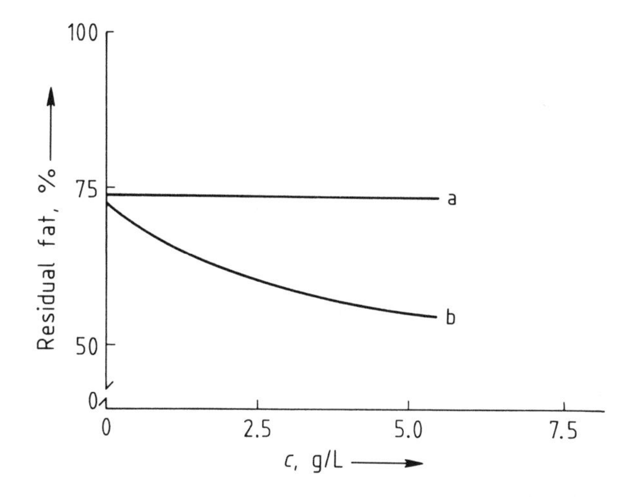

Figure 8. Removal of sebum from polyester/cotton blend fabric studied as a function of electrolyte concentration
Sebum load: 12 g/m^2; apparatus: Launder-ometer; bath ratio: 1:30; wash time: 30 min; temperature: 40 °C
a) Na_2SO_4; b) $Na_5P_3O_{10}$

Removal of the greasy soil by both water (the water value) and sodium sulfate solution is ca. 25 %. Sodium triphosphate increases the value to ca. 45 %. This clearly demonstrates the supplemental washing effect obtained when a complexing agent is introduced to reduce the interfacial tension.

2.3.2 Pigment Soil

Fundamental Principles of Adhesion and Displacement. A theoretical understanding of the interactive forces causing a solid particle to adhere to a more or less smooth surface begins with the Derjaguin–Landau–Verwey–Overbeek theory (DLVO) [26], [27]. Since this theory was developed to explain the phenomena of flocculation and coagulation, however, it can be applied to the washing process only in modified form [28]. A plot of potential energy as a function of the distance of a particle from a fabric shows that the potential energy passes through a maximum (Fig. 9).

The minimum in the potential energy curve corresponds to the closest possible approach, i.e., to the minimum distance *d* that can be established between the pigment and the fibers. The maximum is a measure of the potential barrier that must be overcome if the particle either is removed from the fibers or approaches the fibers from a distance. Adhering particles are more easily removed if the potential barrier is small. Conversely, a soil particle already in the wash liquor is less likely to establish renewed contact with the fabric if the potential barrier is large.

With respect to the washing process, one must recognize that if a particle is bound to a fabric, only a single common electrical double layer located at the overall external surface exists initially. None is present within the zone of contact. During the washing process, new diffuse double layers are created, which cause a reduction

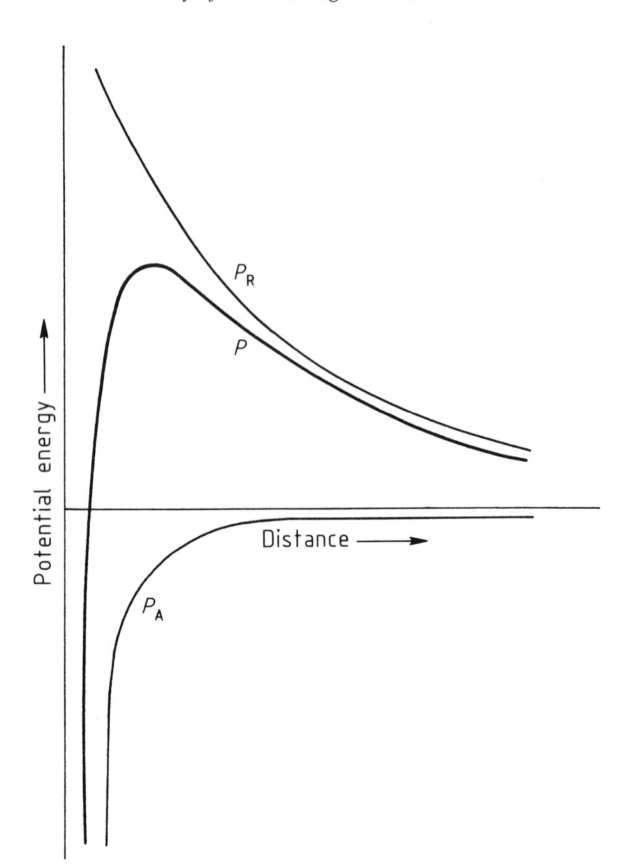

Figure 9. Calculated potential energy of attraction P_A and repulsion P_R as a function of the distance of a particle from a fabric, along with the resultant potential P; predictions based on the DLVO theory DLVO computational parameter $z = 4$

in the free energy of the system. The free energy of an electrical double layer is a function of distance and diminishes asymptotically to a limiting value that corresponds to a condition of no interaction between two double layers. Twice as much effort must be expended to bring a particle into contact with a substrate because of the presence on both surfaces of a double layer (curve $2F$ in Fig. 10). The separation of two adhering surfaces is characterized initially only by van der Waals–London attractive forces P_A and Born repulsion forces P_B, since at this point no electrical double layer exists [28]. In Figure 10, the equilibrium condition corresponding to the potential energy minimum has been taken as the zero point on the abscissa. With increasing distance between the particle and the contact surface, a diffuse double layer arises, which assists in the separation process by establishing an element of repulsion. Thus, the true potential curve P for the separation of an adhering particle in an electrolyte solution results from a combination of the van der Waals–Born potential and the free energy of formation of the electrical double layer. Figures 9 and 10 lead to the important conclusion that an increase in the potential of the

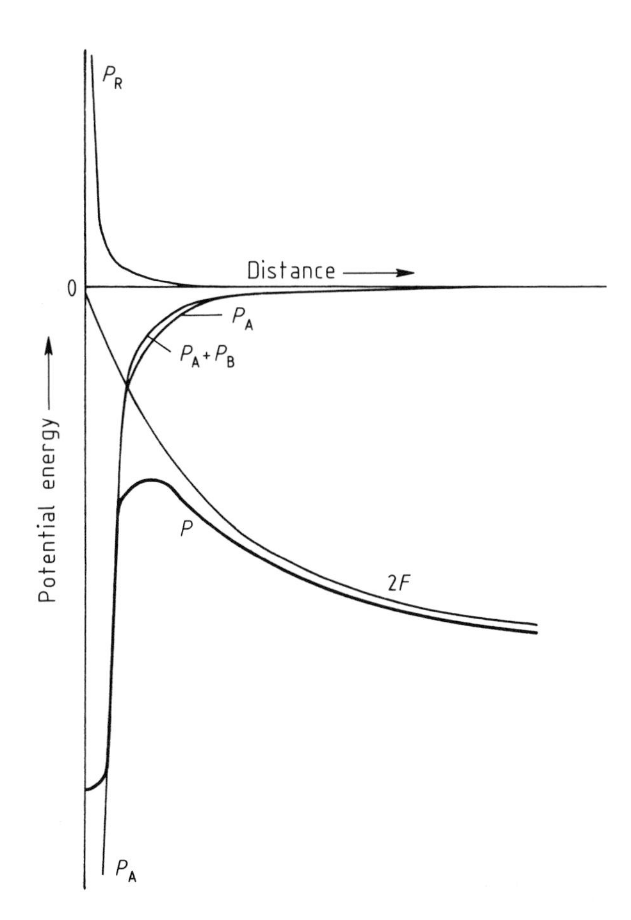

Figure 10. Potential energy diagram for the removal of an adhering particle [28]
P_B Born repulsion; P_A van der Waals attraction; F Free energy of the double layer; P Resulting potential curve
DLVO computational parameter $z = 4$

electrical double layer increases the energy barrier for particle deposition but decreases that for particle removal.

The negative influence exerted by calcium ions derived from water can also be explained with the help of potential theory. According to the Schulze–Hardy rule, compression of an electrical double layer increases rapidly as the valence of a cation increases. Therefore, high concentrations of calcium might cause attractive forces to become the dominant factor, leading to significantly lower washing effectiveness than would be achieved in distilled water.

Effect of Electrical Charges. The foregoing theoretical treatment offers an explanation for the behavior that is actually observed. Surface potentials cannot be measured directly. Instead, the ζ-potential or electrophoretic mobility of a particle is used as a measure of surface charge. As a rule, fibers and pigments in an aqueous medium acquire negative charges, whereby the extent of charge increases with in-

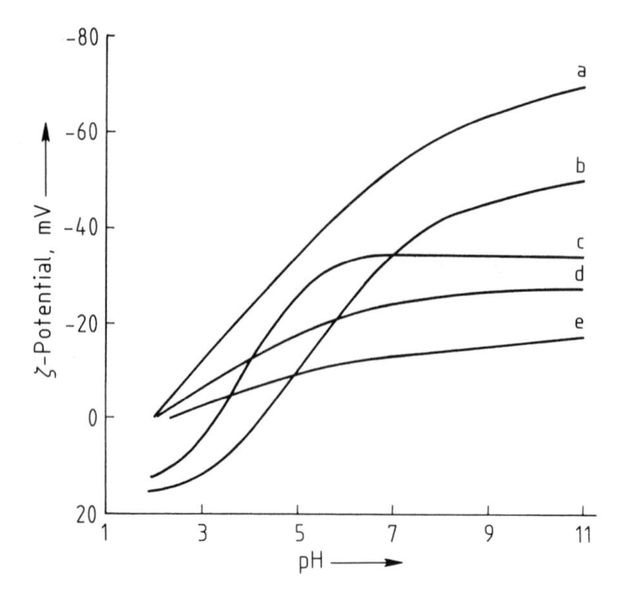

Figure 11. ζ-Potential of various fibers as a function of pH [29]
a) Wool; b) Nylon; c) Silk; d) Cotton; e) Viscose

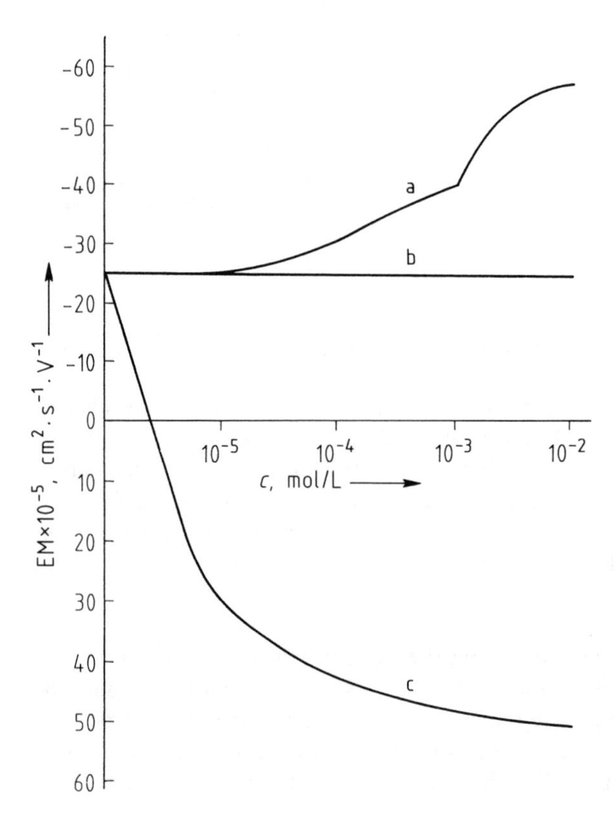

Figure 12. Electrophoretic mobility (EM) of carbon black in solutions containing various surfactants at 35 °C [30]
a) $C_{14}H_{29}OSO_3Na$ (anionic);
b) $C_{14}H_{29}O(CH_2CH_2O)_9H$ (nonionic);
c) $C_{14}H_{29}N(CH_3)_3Cl$ (cationic)

creasing pH. This is illustrated in Figure 11, in which the ζ-potential of various fibers is taken as a measure of electrical charge and is plotted against pH [29].

Essentially similar results are obtained for all major pigment soil components. This is one of the reasons for enhancement of wash performance by the simple introduction of alkali. However, repulsive forces between soil and fibers alone are insufficient to produce satisfactory washing even at high pH.

Apart from changing pH, another way to significantly alter fiber and pigment surface charges is to introduce a surfactant. The sign of the resulting change depends on the nature of the hydrophilic group of the surfactant. Figure 12 shows the effect of several aqueous surfactant solutions on the surface potential of carbon black [30]. The surfactants chosen for the study all share the same hydrophobic component, but their hydrophilic groups vary. Electrophoretic mobility (EM) is taken as a measure of surface potential in this case.

Carbon black also acquires a negative charge in water. The negative charge of pigments and fibers is further increased by adsorption of anionic surfactants. The corresponding increase in mutual repulsion is responsible for an increase in the washing effect. Dispersing power for pigments also increases for the same reason, whereas the redeposition tendency of removed soil is diminished.

In contrast to anionic surfactants, cationic surfactants reduce the magnitude of any negative surface charge. Consequently, electrical repulsion between soil and fabric is lowered. For this reason, cationic surfactants can cause a decrease in washing effect below that observed with pure water that is devoid of additives. Significant soil removal then occurs only at high surfactant concentrations, at which a complete charge reversal takes place on both fabric and soil. When the laundry is

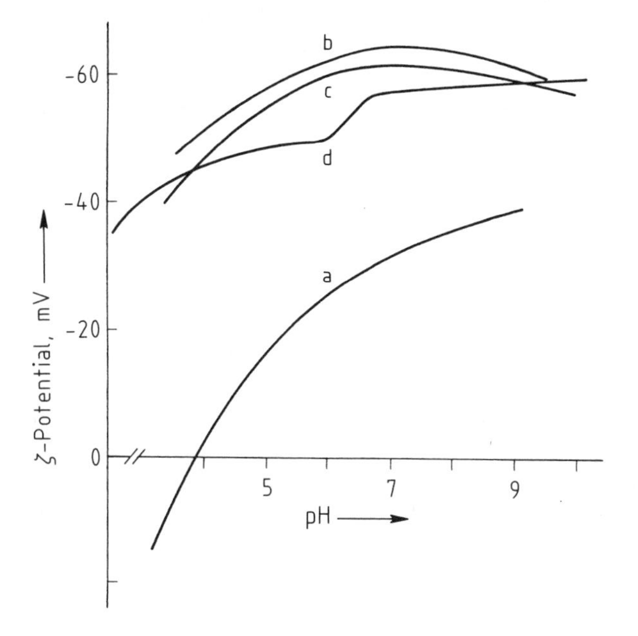

Figure 13. ζ-Potential of hematite as a function of pH at 25 °C in the presence of sodium chloride (a), sodium triphosphate (b), benzenehexacarboxylic acid (c), and 1-hydroxyethane-1,1-diphosphonic acid (d) [7]
Anion concentration: 2.5×10^{-3} mol/L

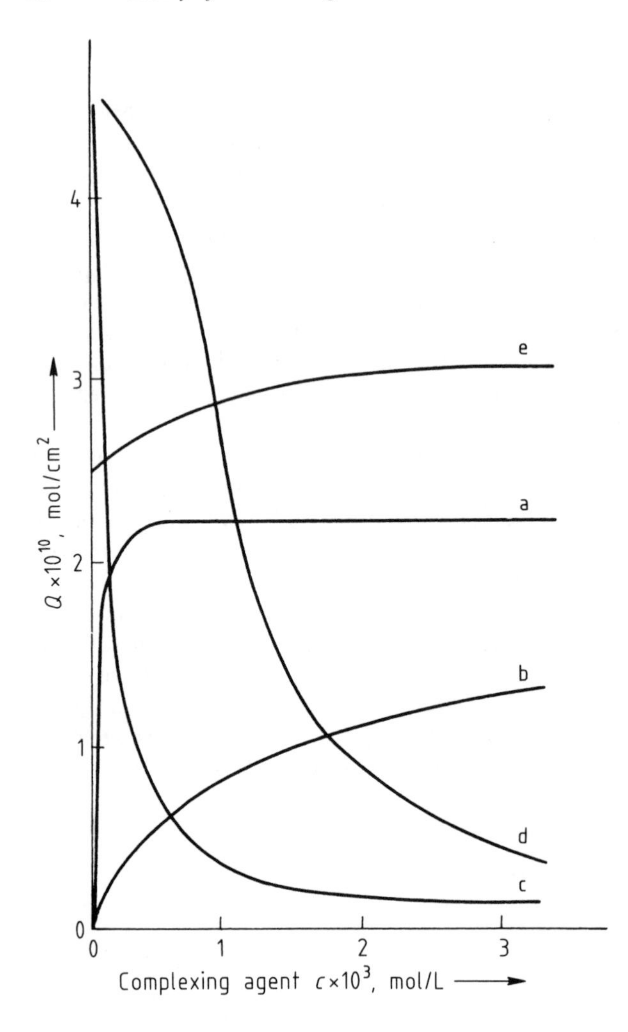

Figure 14. Competitive adsorption Q of complexing agents (sodium triphosphate and sodium citrate) and anionic surfactants on carbon black and γ-aluminum oxide (pH 7, 70 °C) [7]
Surfactant concentration: 1.5×10^{-2} mol/L
a) Adsorption of sodium triphosphate on γ-Al_2O_3;
b) Adsorption of sodium citrate on γ-Al_2O_3;
c) Desorption from γ-Al_2O_3 of the previously adsorbed surfactant *n*-decylbenzenesulfonate by sodium triphosphate;
d) Desorption from γ-Al_2O_3 of the previously adsorbed surfactant *n*-decylbenzenesulfonate by sodium citrate;
e) Adsorption on carbon black of sodium *n*-dodecyl sulfate in competition with sodium triphosphate

rinsed, however, a second charge reversal takes place on the fabric. At the point of electrical neutrality, redeposition of previously removed pigment soil is observed. This is why cationic surfactants are less suited for use in detergents than anionic surfactants.

Whereas anionic and nonionic surfactants are adsorbed nonspecifically at all hydrophobic surfaces, complexing agents can undergo specific attraction to surfaces that have distinct localized charges. The main process is chemisorption and is especially characteristic of metal oxides and certain fibers [7], [31]. As shown in Figure 13, the adsorption of a complexing agent produces an effect similar to that of an anionic surfactant. The change in ζ-potential for hematite is taken as illustrative.

The specificity of complexing agent adsorption with respect to metal oxides is so great that even displacement from boundary surfaces of anionic surfactants with lower adsorption energies is permitted [7]. Figure 14 shows that a poor complexing agent like sodium citrate (curve b) is adsorbed to a lesser extent than sodium triphosphate (curve a) and, thus, displaces correspondingly less anionic surfactant from the boundary surface (curves c and d). Thus, the poorer washing effectiveness of sodium citrate relative to that of sodium triphosphate is a direct consequence of its weaker adsorption.

Although complexing agents suppress the adsorption of anionic surfactants on metal oxides, adsorption is enhanced with materials such as carbon black or synthetic fibers (Fig. 14, curve e). This effect is due to the electrolyte character of the complexing agent. The washing process generally involves removing mixed soils that consist of both hydrophilic and hydrophobic substances from fiber surfaces. For this reason, the differing specificities of complexing agents and surfactants give these two classes of material complementary functions.

Adsorption Layer. In contrast to anionic and cationic surfactants, nonionic surfactants have virtually no effect on surface charge. Therefore, their mode of action cannot be ascribed to changes in electrical charge on pigments and fibers. Instead, the effect is related exclusively to properties of the adsorption layer. Compared to anionic surfactants, the adsorption of nonionic surfactants is quite strong at hydrophobic, weakly polar boundary surfaces. This is illustrated in Figure 15 by the relative adsorptions on activated charcoal of dodecyl dodecaglycol ether and *n*-dodecyl sulfate [30].

In the case of *n*-dodecyl sulfate, both the hydrophilic surfactant groups and the surface have charges of the same sign. Consequently, a higher potential barrier must be overcome for its adsorption compared to the electrically neutral nonionic surfactant. With nonionic surfactants, both the adsorption equilibrium and the maximum surface coverage are displaced to relatively low concentration. Figure 16 is a sche-

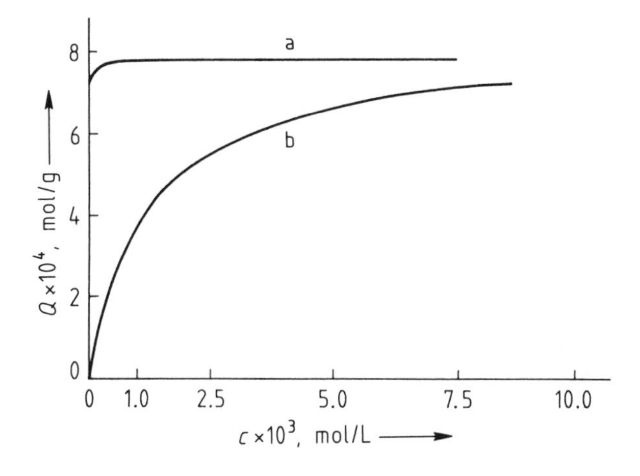

Figure 15. Adsorption isotherms for surfactants on activated charcoal [30]
25 °C, surface 1150 m^2/g (from BET measurements)
a) $C_{12}H_{25}O(CH_2CH_2O)_{12}H$;
b) $C_{12}H_{25}OSO_3Na$

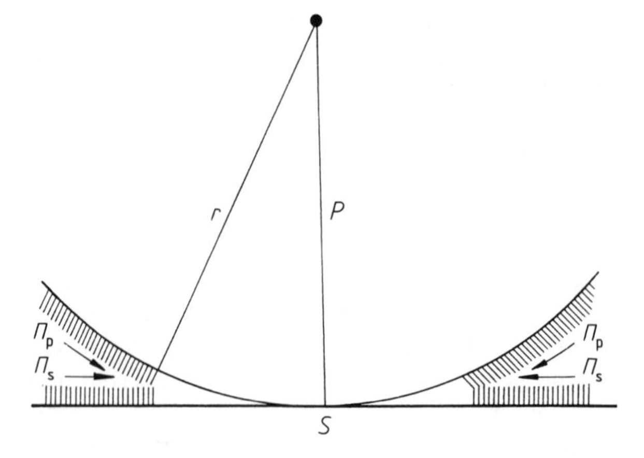

Figure 16. Schematic representation of adsorption-induced separation of a spherical particle from a hard surface
S Surface; P Particle;
Π_S Splitting pressure of the surfactant layer on the surface; Π_P Splitting pressure of the surfactant layer on the particle

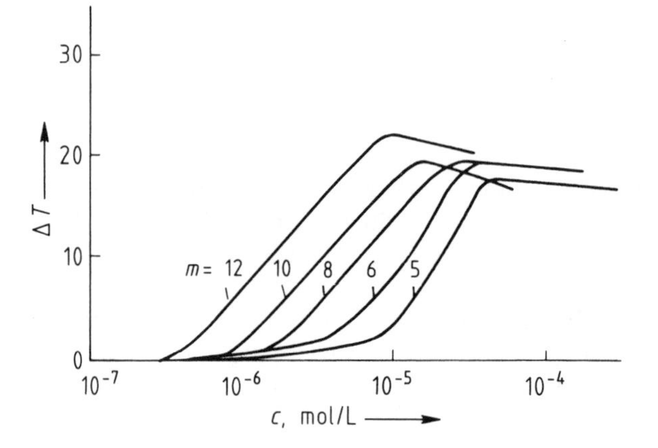

Figure 17. Reduction in the turbidity ΔT of a paraffin sol by addition of dodecyl polyglycol ethers ($C_{12}H_{25}O(CH_2CH_2O)_mH$) in the presence of 0.5 mol/L NaCl [32]

matic representation of the adsorption layers on substrate and soil particles. One can see from the diagram that both surfactant layers advance to the point of pigment–surface contact.

One consequence is the development of a splitting pressure, which leads to separation of the soil particle from the surface. This effect is obviously present with anionic surfactants as well. However, this pressure is the decisive factor with nonionic surfactants, due to the absence of any repulsive components of electrostatic origin.

Thus, hydration of hydrophilic groups is extremely important with nonionic surfactants. Adsorbed surfactant molecules are so oriented that their hydrophilic regions are directed toward the aqueous phase. Both pigment and substrate are surrounded by hydration spheres. The redeposition tendency of a soil particle is thereby reduced, because the voluminous hydration sphere minimizes the effectiveness of short-range van der Waals attractive forces. This can be demonstrated by the coagulation of hydrophobic sols in the presence of nonionic surfactants containing varying numbers of ethoxy groups in their hydrophilic regions [32].

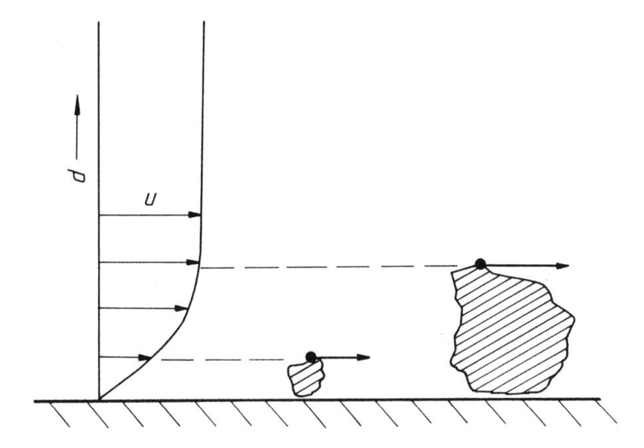

Figure 18. Schematic drawing of the effect of liquid flow on adhering particles of different size
u Flow velocity; *d* Distance from the solid surface

Figure 17 illustrates the stabilizing effects of dodecyl polyglycol ethers on a paraffin oil sol, where reduction in turbidity, ΔT, of the coagulated sol is plotted against the concentrations of ethers containing varying numbers of ethoxy groups. Coagulation is induced in this case by the presence of large amounts of sodium chloride, which causes compression of the electrical double layer. The turbidity decrease becomes greater as the number of ethoxy groups in the molecule rises, and the curves are displaced in the direction of lower concentration.

Hydrodynamics. In practice, the physico-chemical principles discussed thus far are augmented by taking advantage of hydrodynamic effects. Such hydrodynamic effects generally are strongly dependent on particle size [33]; their significance in the removal of pigment soil from fibers increases as particle size increases. Even with vigorous mechanical action, a laminar film of finite thickness in which no flow takes place is present on every surface. The flow velocity gradient increases with increasing distance from the surface, as shown schematically in Figure 18.

Thus, an increase in mechanical force has a significant soil-removing effect on larger particles. A high flow velocity gradient in the immediate vicinity of a surface is excluded on hydro-dynamic grounds. For this reason, washing machines employ abrupt changes in direction of operation to achieve adequate turbulence near substrate surfaces. Even so, particles smaller than 0.1 μm cannot be displaced by mechanical means alone.

2.3.3 Calcium-Containing Soil

The principles discussed in Section 2.3.2 also apply without exception to calcium-containing pigment soil. However, other important mechanisms, which deserve discussion, also exist. Salts of multivalent cations are almost always present in soils and

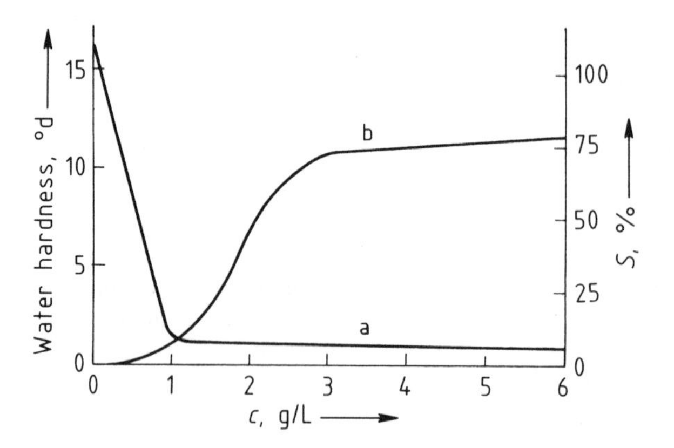

Figure 19. Water softening (a) and soil removal *S* from cotton soiled with a dust/sebum mixture (b) resulting from a cross-linked water-insoluble polyacrylate at 90 °C [7]

on textile fiber surfaces. Salts such as calcium carbonate, calcium phosphate, and calcium stearate are especially prevalent. However, cationic bridges, which are responsible for binding soil components chemically to fibers, also frequently form. This type of linkage can be due, for example, to the carboxyl groups commonly found in cotton as a result of oxidation, but such a linkage may also arise from the presence of reactive centers associated with metal oxides or from soaps derived from sebum. Problems stem principally from poorly soluble calcium salts, whose solubility is further diminished as the water hardness of the wash liquor increases. On the other hand, their solubility in distilled water is higher because of displacement of the solubility equilibrium. When calcium salts are dissolved from a multilayered soil deposit, "holes" remain in the structure. These holes loosen the deposit and facilitate its removal from the surface [11]. Thus, one task of detergent components is to create the highest possible calcium ion concentration gradient between soil and aqueous phase during the washing process.

Figure 19 illustrates this principle by demonstrating the wash effectiveness of a water-insoluble cross-linked polyacrylate [7], [34]. Apparently, virtually no wash effectiveness is achieved at a concentration of ion exchanger sufficient to eliminate most of the water hardness; an effect is observed only at a higher concentration of ion exchanger. This phenomenon can be greatly accelerated by also introducing an appropriate amount of water-soluble complexing agent.

Before the slightly soluble cations can be dissolved out of the soil and fibers, adsorption of the complexing agent takes place on the surface, particularly in those areas containing multivalent cations. In the course of subsequent desorption of water-soluble multivalent cation complexes, many of the soil–fiber bonds are broken, leading to a marked enhancement of the washing effect. Removal of cations from soil and fibers by adsorption–desorption processes and displacement of solubility equilibria are the most important phenomena that accompany the use of complexing agents and ion exchangers in the washing process.

The mechanisms of action for complexing agents and water-insoluble ion exchangers differ; thus, the two complement each other in their respective roles [35]. Figure 20 shows the way in which a small amount of a water-soluble complexing

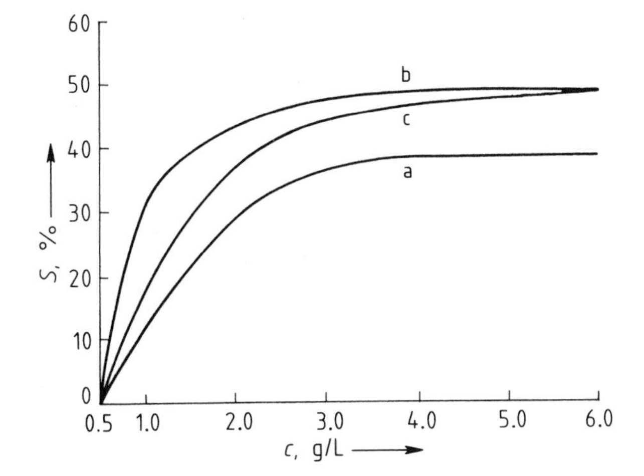

Figure 20. Comparison of soil removal *S* of zeolite 4 A (a), sodium triphosphate (b), and a mixture of the two builders in the ratio zeolite 4 A: sodium triphosphate = 9 : 1 (c); results obtained for non-resin-finished cotton tested in a Launderometer [35]
Wash time: 30 min with heating; temperature: 90 °C; water hardness: 16 °d

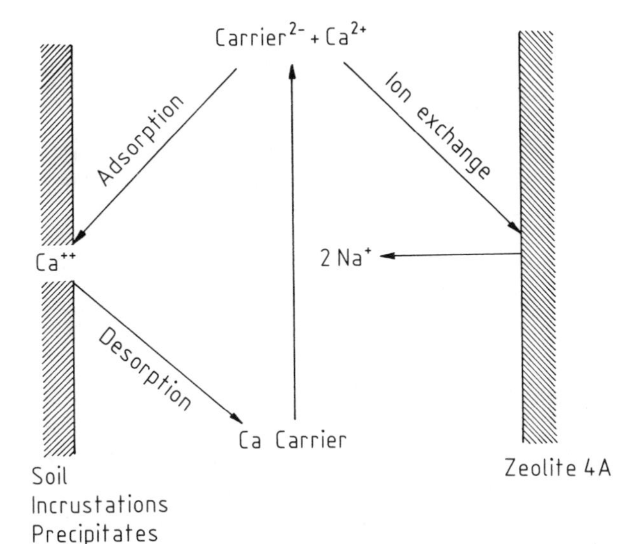

Figure 21. Mechanistic scheme for the carrier effect [29]

agent can increase the washing effectiveness of the water-insoluble ion exchanger zeolite 4 A. Zeolite 4 A – sodium triphosphate or zeolite 4 A detergents are currently experiencing increasing commercial success alongside detergents containing primarily sodium triphosphate. The effect shown in Figure 20 results from an increase in the rate of dissolution of divalent ions out of the soil and fibers [36]. The mode of action is depicted schematically in Figure 21. The water-soluble complexing agent serves as a carrier that transports calcium out of the precipitate and into the water-insoluble ion exchanger. The process is based on successive adsorption, desorption, and dissociation, and it accelerates the delivery of free calcium ions into solution.

Table 8. Wash experiments in a calcium-free system [36]

Soil	Wash medium *	Remission, %
80.2% Osmosed kaolin, 16.5% carbon black, 3.3% black iron oxide	H_2O	65.5
	H_2O + 2 g $Na_5P_3O_{10}$/L	66.0
	H_2O + 2 g zeolite 4A/L	66.0
89.7% Osmosed kaolin, 5.9% carbon black, 2.9% black iron oxide, 1.5% yellow iron oxide	H_2O	59.5
	H_2O + 2 g $Na_5P_3O_{10}$/L	58.0
	H_2O + 2 g zeolite 4A/L	59.0

* Water is distilled.

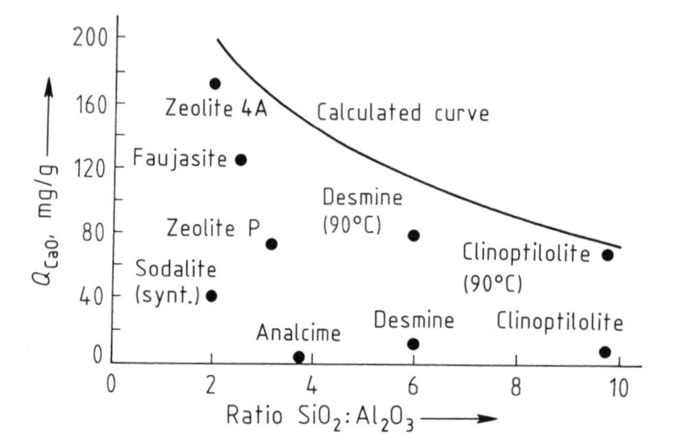

Figure 22. Calcium binding capability Q of various zeolites [37]
Reaction time: 15 min; temperature: 22 ± 1 °C; Ca^{2+} concentration: 30 °d; zeolite concentration: 1 g/L

The effectiveness of complexing agents and ion exchangers is related to the presence of calcium in the system, as is evident from Table 8. Two different soils and the cotton yarn to be studied were decalcified prior to deposition of an artificial soil. No washing effect due either to sodium triphosphate or to zeolite 4A could be observed within the method's limits of accuracy. This result can be taken as indirect proof of the importance of dissolution of calcium from soil and fibers during the washing process.

The concentration of complexing agent and the temperature are generally the decisive factors in removing multivalent metal ions by a water-soluble complexing agent; the binding ability diminishes with increasing temperature.

Essentially the same relationships also apply with a water-insoluble ion exchanger, although the temperature effect is usually reversed, as can be seen for two examples in Figure 22 [37], [38].

The reversed sign of the temperature effect of calcium binding to zeolites relative to water-soluble complexing agents is a major advantage of using zeolite 4A–sodium triphosphate mixtures in heavy-duty detergents.

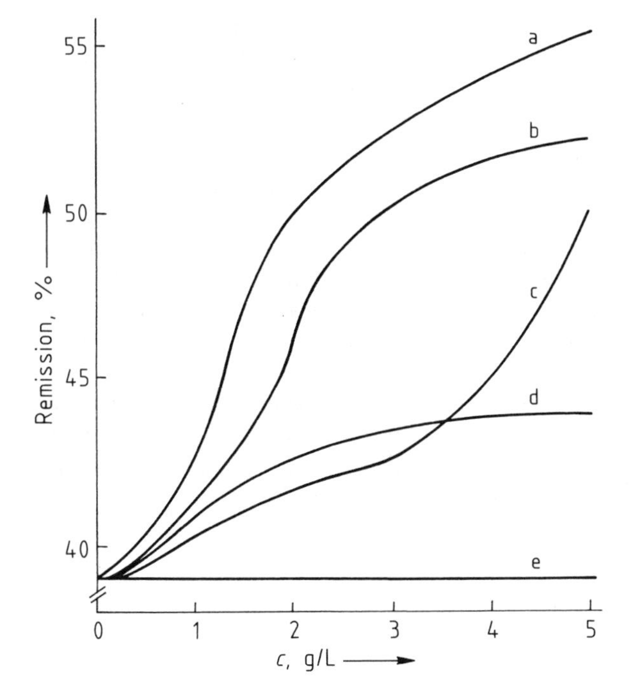

Figure 23. Soil removal of various zeolites [37]
a) Zeolite 4 A; b) Faujasite; c) Desmine; d) Sodalite; e) Analcime
Apparatus: Launder-ometer; water hardness: 16 °d; temperature: 95 °C; wash time: 30 min with heating; fabric: non-resin-finished cotton

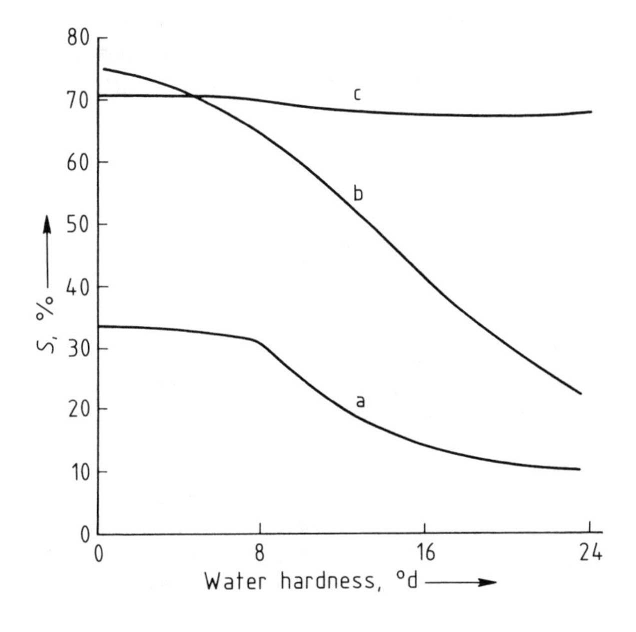

Figure 24. Soil removal S from wool as a function of water hardness at 30 °C [7]
a) Sodium alkylbenzenesulfonate (0.5 g/L); b) Sodium alkylbenzenesulfonate (0.5 g/L) together with sodium sulfate (1.5 g/L); c) Sodium alkylbenzenesulfonate (0.5 g/L) together with sodium triphosphate (1.5 g/L)

Figure 23 shows the wash effectiveness of various zeolites in water with a hardness of 16 °d as a function of concentration. It is apparent that the best results are achieved with zeolite 4 A and the poorest with analcime.

In addition to the previously described reasons for removing divalent alkaline-earth ions, one must also consider their interaction with other detergent components. For example, soaps form poorly soluble salts with calcium, as do many synthetic surfactants, and these can be deposited on fibers. This phenomenon was extremely common with earlier detergents, which were comprised mainly of soap; at that time sufficiently strong complexing agents such as sodium triphosphate had not yet come into use. Precipitation of relatively insoluble surfactant calcium salts has the additional disadvantage that it leads to a severe reduction in active surfactant concentration and, thus, to generally poorer conditions for soil removal.

Figure 24 illustrates how surfactants, salts, and complexing agents complement each other in the removal of soil. The negative influence of calcium is eliminated by sodium triphosphate, and the magnitude of the indirect counterion effect of sodium ions is shown by addition of sodium sulfate to alkylbenzenesulfonate.

2.3.4 Influence of Textile Fiber Type

The ability of a detergent to remove soil depends not only on the foregoing factors, but also on the type of textile substrate. Textile fibers that have a high calcium content at their surface (e.g., cotton) behave very differently from synthetic fibers with a low calcium content. Fiber type has a dramatic influence on the degree of hydrophobicity/hydrophilicity, the wettability, and the extent of soil removal. Figure 25 clearly demonstrates through soil removal the differing effects that complexing agents and surfactants have on a series of fibers. The anionic surfactant selected in this case for study (alkylbenzenesulfonate) has virtually no effect on

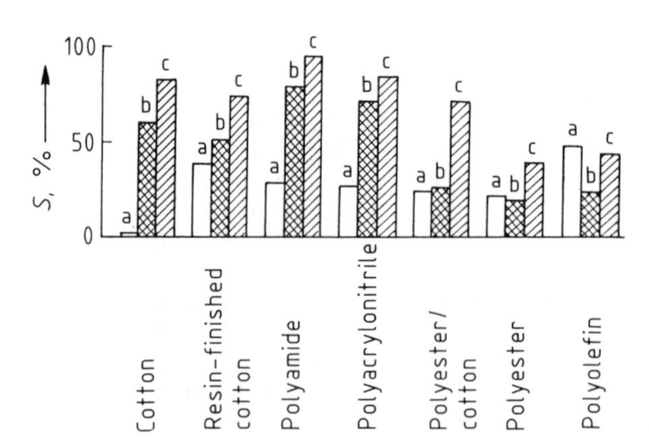

Figure 25. Influence of textile fibers on soil removal S [7]
Detergents: a) 1 g/L Alkylbenzenesulfonate + 2 g/L sodium sulfate; b) 2 g/L sodium triphosphate; c) 1 g/L alkylbenzenesulfonate + 2 g/L sodium triphosphate

cotton, whereas even by itself sodium triphosphate leads to substantially improved wash effectiveness. Resin-finished cotton also shows a greater effect for sodium triphosphate than for surfactants alone. The effect of sodium triphosphate is particularly high with the relatively hydrophilic synthetic fibers polyamide and polyacrylonitrile. This behavior changes when more hydrophobic textile fiber are encountered. An effect is still apparent with polyester/cotton and polyester, but it is not greater than that of surfactants alone. At the same time, overall wash performance is diminished. In the case of the very hydrophobic polyolefin fibers, the wash effectiveness of the surfactant is substantially greater than that of the complexing agent. These examples show the exceptional way in which surfactants and complexing agents or ion exchangers complement each other, not only in the case of mixed soils, but also from the standpoint of various fibers.

2.4 Subsequent Processes

After soil has been removed, it must be stabilized in the wash liquor, and redeposition of the removed soil must be prevented. This property of a wash liquor is known as its *soil antiredeposition capability*. Several mechanisms play a role in ensuring good antiredeposition characteristics.

2.4.1 Dispersion and Solubilization Processes

The most important factor in this context is the dispersion process. Nonspecific adsorption of surfactants and specific adsorption of complexing agents causes liquid soils to be emulsified and solid soils suspended. Poorly soluble substances are solubilized by surfactant micelles as molecular dispersions. These mechanisms have been discussed in detail in Sections 2.3.1 and 2.3.2.

2.4.2 Adsorption

Supplementary adsorption effects play an important role when water-insoluble ion exchangers of the zeolite type are used in detergents. Such effects are not observed with detergents containing only the complexing agent sodium triphosphate.

Zeolites can significantly enhance washing effectiveness by serving as competitive substrates for the adsorption of molecularly dispersed soluble substances and col-

Table 9. Comparison of the washing power of phosphate-containing detergents with that of model detergents containing zeolite 4A; washing machine tests of various kinds of white fabrics in the presence and absence of cloth that releases a black dye [35]

Model detergent	Remission, %					
	Non-resin-finished cotton		Resin-finished cotton		Polyester/cotton	
	without dyed fabric	with black fabric	without dyed fabric	with black fabric	without dyed fabric	with black fabric
With sodium triphosphate	83	67	74	66	70	49
With zeolite 4A–sodium triphosphate	82	75	73	66	74	57

loidal particles. Their presence is particularly advantageous under extreme conditions involving a large amount of pigmented material. Adsorption and heterocoagulation of the soil by zeolites substantially reduce redeposition on the fabric, leading to a significant increase in whitening power.

Table 9 illustrates this phenomenon by comparing the results obtained in a washing machine when various white fabrics were washed in the presence and in the absence of cloth treated with a nonfast black dye. Tests were conducted with phosphate-containing detergents and with model detergents containing zeolites. The effects are clear, pointing up the importance of competition between zeolites and textile substrates for substances prone to cause graying of laundry. If conditions are less extreme, the differences become apparent only after the washing process has been repeated many times, as shown by the data in Table 10. Again, redeposition tendencies are reduced by detergents containing zeolites [20], [36].

2.4.3 Soil Antiredeposition Effect by Polymers

Soils and detergents also contain natural and synthetic macromolecules in addition to low molecular mass compounds. Natural proteins from blood and protein-containing foods can be adsorbed onto textile fibers and must by some means be desorbed during the wash process.

Detergents often contain polymeric soil antiredeposition agents, whose role is to be adsorbed onto the substrate, thereby creating a protective layer that sterically inhibits redeposition of previously removed pigment soil. The desorption of natural proteins and the adsorption of antiredeposition agent in a single process represent competing phenomena, so that careful selection of the proper antiredeposition agent is required.

Table 10. Comparison of changes in polyester wetting tensions with changes in remission for a heavy-duty detergent (I) and a low-temperature heavy-duty detergent (II) for polyester

Detergent I,	Detergent II,	Antiredeposition agent (in % detergent charge)			Change in wetting tension *		Change in remission **	
g/L	g/L	Sodium carboxymethyl cellulose	Methylhydroxypropyl cellulose	Hydroxyethyl cellulose	I Δj_v, mN/m	II Δj_v, mN/m	I ΔR, %	II ΔR, %
7.4	4.5	0	0	0	0	−1		
0	0	0.5	0	0	1	1		
7.4	4.5	0.5	0	0	2	0	3	−1
0	0	0	0.5	0	22	22		
7.4	4.5	0	0.5	0	14	14	17	16
0	0	0	0	0.5	22	21		
7.4	4.5	0	0	0.5	3	2	5	0

* Δj_v = calculated for advancing contact angle conditions. ** ΔR = measured after three washes.

In a surfactant-free environment, the adsorption of most macromolecules onto a solid surface is effectively an irreversible process. The reason for this irreversibility is the vast number of points of contact that exist between macromolecule and substrate. From a statistical standpoint, the number of bonds between the two is so large that adherence is ensured regardless of the strength of any single attachment. Most of the observed binding results from weak hydrophobic interactions.

In a multicomponent system containing both surfactants and macromolecules, competitive adsorption is possible at the substrate surface. This permits the surfactant to successively destroy the individual points of contact binding the macromolecule, thereby displacing it from the boundary surface. This mechanism is most often observed for nonionic surfactants. Anionic surfactants are often capable of forming polymer–surfactant complexes, in the course of which the conformation of the macromolecule is altered in such a way as to reduce the extent of its attraction to the boundary surface.

Figure 26 depicts the adsorption behavior of gelatine on glass both in the presence and in the absence of added sodium *n*-dodecyl sulfate. The two adsorption isotherms are very different; in the presence of a surfactant, adsorption of the macromolecule is virtually eliminated. Even preadsorbed gelatine is desorbed from the surface by subsequent surfactant addition. This competitive adsorption phenomenon involving surfactants and macromolecules is of great importance in soil removal, as is the formation of macromolecule–surfactant complexes. Both significantly impair the desired adsorption of polymeric antiredeposition agents.

The adsorption of antiredeposition agents is usually a selective process dependent on the chemical constitutions of both substrate and polymer. For example, the soil antiredeposition effect of carboxymethyl cellulose is rather limited on hydrophilic fibers such as cotton. Cellulose ethers, such as methylhydroxypropyl cellulose, are effective not only with hydrophilic fibers, but also especially with more hydrophobic fibers such as polyester. Therefore, combinations of several antiredeposition agents often must be used to ensure satisfactory results with mixed laundry. In this case the absolute amount of adsorbed substance is not the determining factor, but

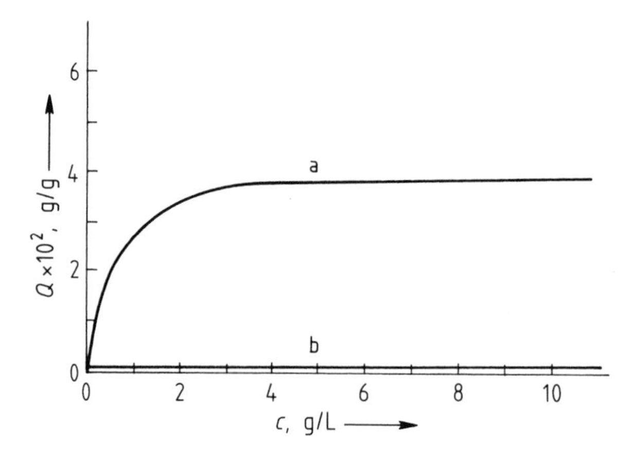

Figure 26. Adsorption of gelatine on powdered glass (temperature 25 °C)
a) Without sodium dodecyl sulfate; b) With sodium dodecyl sulfate (gelatine–surfactant mass ratio 1:1.44)

rather the extent to which adsorption confers hydrophilic characteristics, i.e., the change in surface characteristics relative to untreated fiber surfaces. This can be characterized by observing the resulting differences in the wetting tension with respect to pure water (cf. Eq. 1 and 2). Table 10 shows that carboxymethyl cellulose, long used as an antiredeposition agent for cotton, has no effect whatsoever on polyester. By contrast, methylhydroxypropyl cellulose causes the polyester surface to become considerably more hydrophilic. Noteworthy is the fact that the effects are retained with both detergents, albeit to a somewhat reduced extent. Only with these formulations does one observe a significant increase in the soil antiredeposition effect, as evidenced by the changes in remission. Hydroxyethyl cellulose can be regarded as a representative of numerous polymers which, though they are readily adsorbed out of an aqueous solution and are capable of showing considerable antiredeposition activity in pure water, nonetheless lose most of their effectiveness in a detergent solution as a result of competitive adsorption and displacement by surfactants.

2.5 Concluding Remarks

To simplify the theoretical treatment of the washing process, it has been necessary to treat separately each of the several phenomena involved and to isolate them from one another. In any real washing process, the various mechanisms are all at work more or less simultaneously. Thus, these mechanisms affect one another in a mutually supportive and additive way. Investigations into the theory of washing have now made possible a rather thorough understanding of the process. Despite the complexity of the washing phenomenon and the continued presence of certain unanswered questions, these physicochemical correlations have exerted a major influence on product development.

3 Detergent Ingredients

Detergents for household and institutional use are very complex formulations containing several different types of substances. These can be categorized into the following major groups:

surfactants
builders
bleaching agents
auxiliary agents

Each individual component of a detergent has its own very specific functions in the washing process, although to some extent they have synergistic effects on one another. In addition to the above substances, certain additives are made necessary by the production process, whereas other materials are introduced to improve product appearance.

3.1 Surfactants

Surfactants constitute the most important group of detergent components, and they are present in all types of detergents. Generally, these are water-soluble surface-active agents comprised of a hydrophobic portion (usually a long alkyl chain) attached to hydrophilic or solubility-enhancing functional groups.

A surfactant can be placed in one of four classes, depending on what charge is present in the chain-carrying portion of the molecule after dissociation in aqueous solution:

anionic surfactants
nonionic surfactants
cationic surfactants
amphoteric surfactants

Table 11 provides an overview of these classes.

In general, both adsorption and wash effectiveness increase with increasing chain length. For example, ionic surfactants bearing *n*-alkyl groups show a linear relation-

Table 11. Surfactants of various ionic nature [39]

Surfactant	Formula	Electrolytic dissociation	Ionic nature
Alkyl poly(ethylene glycol) ethers	$RO-(CH_2-CH_2-O)_nH$	no	nonionic
Alkylsulfonates	$R-SO_3^- \; Na^+$	yes	anionic
Dialkyldimethylammonium chlorides	$\left[H_3C-\overset{R}{\underset{R}{\overset{\vert}{\underset{\vert}{N}}}}{}^{+}-CH_3\right] Cl^-$	yes	cationic
Betaines	$R-\overset{CH_3}{\underset{CH_3}{\overset{\vert}{\underset{\vert}{N}}}}{}^{+}-CH_2-C\genfrac{}{}{0pt}{}{=O}{-O^-}$		amphoteric

ship between the number of carbon atoms in the surfactant molecule and the logarithm of the amount of surfactant adsorbed on activated carbon or kaolin [40], [41].

The structure of the hydrophobic residue also has a significant effect on surfactant properties. Surfactants with little branching in their alkyl chains generally show good wash effectiveness but relatively poor wetting characteristics, whereas more highly branched surfactants are good wetting agents but have unsatisfactory detergency. For compounds containing an equal number of carbon atoms in their hydrophobic residues, wetting power increases markedly as the hydrophilic groups move to the center of the chain or as branching increases, but a simultaneous decrease in adsorption and washing power occurs (Figs. 27 and 28).

The changes with respect to adsorption, wetting, and wash effectiveness that result from varying the degree of branching are far more significant for ionic surfactants than for non-ionic surfactants. In the case of anionic surfactants, losses in wash effectiveness caused by increased branching can be recovered to some extent, provided the overall number of carbon atoms is increased in an appropriate fashion.

Household washing of textiles normally poses few situations requiring extraordinary wetting power. If problems do arise, they can usually be overcome by increasing the wash time or the amount of detergent used. Most important is the effectiveness of the rolling-up process. Optical microscopy has revealed that oily and greasy soil tends to reside in more or less evenly distributed layers on the surface of fibers. These layers are gradually constricted by the action of a surfactant and its associated spreading pressure, with the layers ultimately being reduced to drops resting loosely on the fibers, which are then easily rinsed off into the wash liquor. Thus, oily residue as a fiber wetting agent is replaced by aqueous wash solution, and the process by which it occurs has come to be known as *roll-up*.

The number of types of surfactant suitable for use in detergents has increased considerably in the past 30 years. The principal criteria for judging surfactant suit-

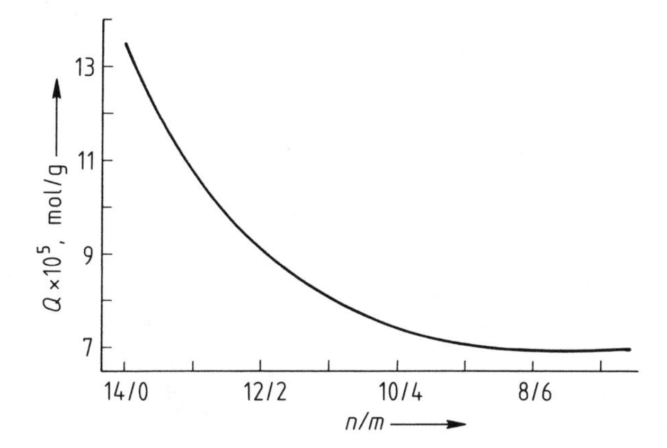

Figure 27. Decrease in adsorbed amount Q at equilibrium with increased branching of the hydrophobic residue [40]
Adsorbent: activated carbon M; amount of adsorbent: 0.050 g; particle diameter: 0.084 cm; surfactant:

$$\begin{array}{l} C_nH_{2n+1} \\ \quad \backslash \\ \quad CH{-}CH_2{-}OSO_3Na \\ \quad / \\ C_mH_{2m+1} \end{array}$$

surfactant concentration: 1×10^{-4} mol/L; temperature: 25 °C

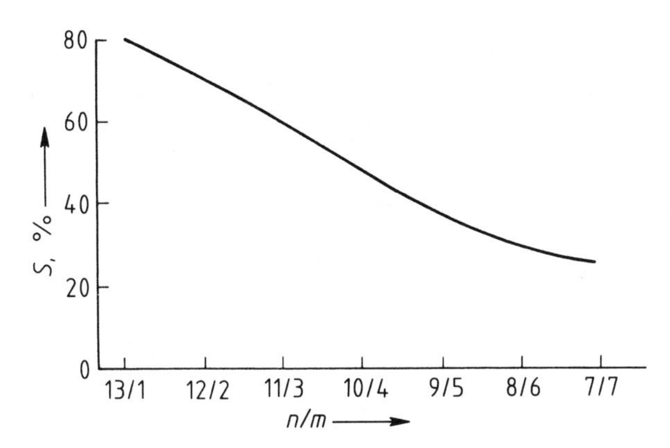

Figure 28. Decrease in soil removal S from soiled cotton as a function of increased branching in the hydrophobic residue [40]
Surfactant:

$$\begin{array}{l} C_nH_{2n+1} \\ \quad \backslash \\ \quad CH{-}CH_2{-}OSO_3Na \\ \quad / \\ C_mH_{2m+1} \end{array}$$

temperature: 90 °C; bath ratio: 1:12.5; water hardness: 16 °d; surfactant concentration: 2.91×10^{-3} mol/L

Table 12. Key surfactants

Structure		Chemical name	Acronym
Anionic surfactants			
$R-CH_2-C(=O)-ONa$	$R = C_{10-16}$	soaps	
$R-C_6H_4-SO_3Na$	$R = C_{10-13}$	alkylbenzenesulfonates	LAS
$(R^1)(R^2)CH-SO_3Na$	$R^1 + R^2 = C_{11-17}$	alkanesulfonates	SAS
$H_3C-(CH_2)_m-CH=CH-(CH_2)_n-SO_3Na$ + $R-CH_2-CH(OH)-(CH_2)_x-SO_3Na$	$n + m = 9-15$; $n = 0,1,2...$ $m = 1,2,3...$; $R = C_{7-13}$ $x = 1,2,3$	α-olefinsulfonates	AOS
$R-CH(SO_3Na)-C(=O)-OCH_3$	$R = C_{14-16}$	α-sulfo fatty acid methyl esters	SES
$R-CH_2-O-SO_3Na$	$R = C_{11-17}$	fatty alcohol sulfates, alkyl sulfates	FAS
$R^2-CH(R^1)-CH_2-O-(CH_2-CH_2-O)_n-SO_3Na$	a) $R^1 = H$, $R^2 = C_{10-12}$; b) $R^1 + R^2 = C_{11-13}$, $R^1 = H, C_1, C_2...$; $n = 1-4$	alkyl ether sulfates a) fatty alcohol ether sulfates b) oxo alcohol ether sulfates	FES

Table 12. (continued)

Structure		Chemical name	Acronym
Cationic surfactants			
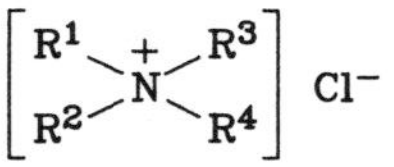 $\left[R^1R^2N^+R^3R^4 \right] Cl^-$	$R^1, R^2 = C_{16-18}$ $R^3, R^4 = C_1$	quaternary ammonium compounds tetraalkylammonium chloride	QAC
Nonionic surfactants			
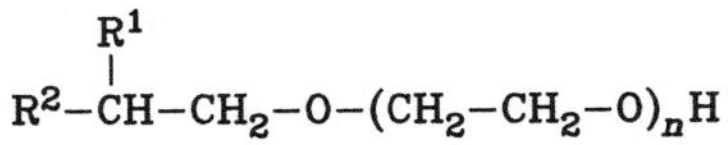 $R^2-CH(R^1)-CH_2-O-(CH_2-CH_2-O)_nH$	a) $R^1 = H \quad R^2 = C_{6-16}$ b) $R^1 + R^2 = C_{7-13}$ $R^1 = H, C_1, C_2 \ldots$ $n = 3-15$	alkyl poly(ethylene glycol) ethers a) fatty alcohol poly(ethylene glycol) ethers b) oxo alcohol poly(ethylene glycol) ethers	AEO
$R-C_6H_4-O-(CH_2-CH_2-O)_nH$	$R = C_{8-12} \quad n = 5-10$	alkylphenol poly(ethylene glycol) ethers	APEO
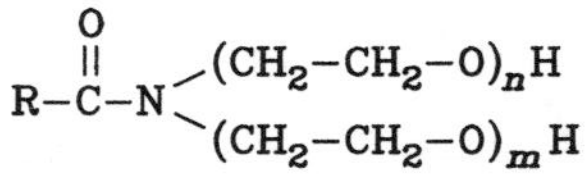 $R-C(=O)-N[(CH_2-CH_2-O)_nH][(CH_2-CH_2-O)_mH]$	$R = C_{11-17}$ $n = 1, 2 \quad m = 0, 1$	fatty acid alkanolamides	FAA
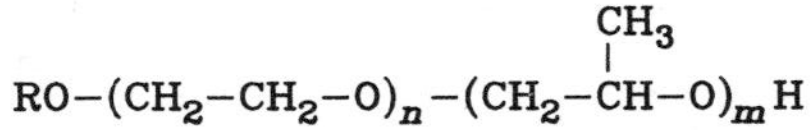 $RO-(CH_2-CH_2-O)_n-(CH_2-CH(CH_3)-O)_mH$	$R = C_{8-18}$ $n = 3-6 \quad m = 3-6$	fatty alcohol polyglycol ethers (EO/PO adducts)	FEP

Table 12. (continued)

Structure		Chemical name	Acronym
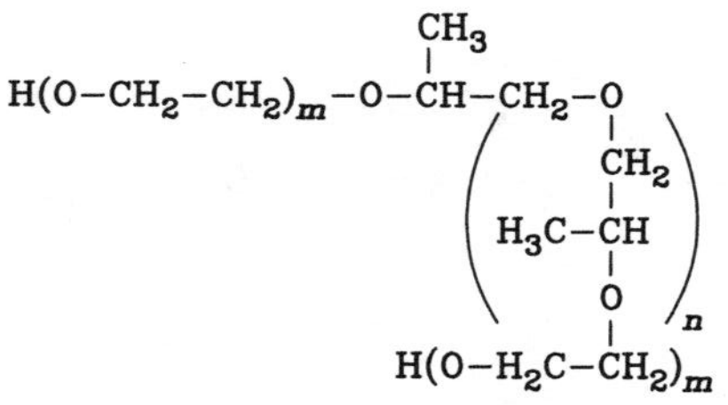 $H(O{-}CH_2{-}CH_2)_m{-}O{-}CH(CH_3){-}CH_2{-}O{-}\left(CH_2{-}CH(H_3C){-}O\right)_n{-}(CH_2{-}CH_2{-}O)_m H$	n = 2-60 m = 15-80	ethylene oxide–propylene oxide block polymers	EPE
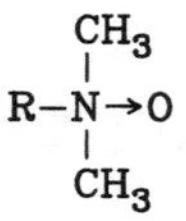 $R{-}N(CH_3)_2{\rightarrow}O$	$R = C_{12-18}$	alkyldimethylamine oxides	
Amphoteric surfactants			
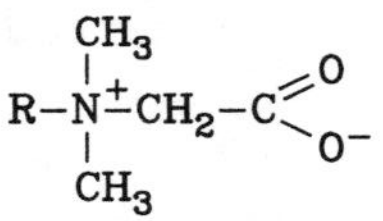 $R{-}N^{\pm}(CH_3)_2{-}CH_2{-}C({=}O){-}O^-$	$R = C_{12-18}$	alkylbetaines	
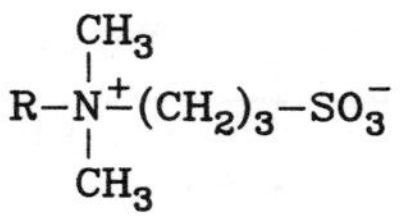 $R{-}N^{\pm}(CH_3)_2{-}(CH_2)_3{-}SO_3^-$	$R = C_{12-18}$	alkylsulfobetaines	

ability apart from performance are toxicological and ecological characteristics. Cationic and nonionic surfactants have come to play an increasingly important role along with their anionic counterparts (cf. Table 12). Despite the wide choice of possibilities, however, only a very few surfactants account for the major share of the market, partly as a result of economic factors.

Anionic surfactants are the most common agents in detergents designed for laundry, dishwashing, and general cleansing, although non-ionic surfactants of the ethylene oxide adduct variety have also acquired great importance. Cationic surfactant use is largely restricted to aftertreatment aids because of the fundamental incompatibility of these materials with anionic surfactants. Amphoteric surfactants still lack a significant place in the market. Around the world a remarkable variability in the types and amounts of surfactants employed in products for similar purposes can be seen. The reasons are to be found in the variations in the kinds of fabric encountered worldwide, the diversity in washing machine technology, and different national customs for fabric use and care (Fig. 29).

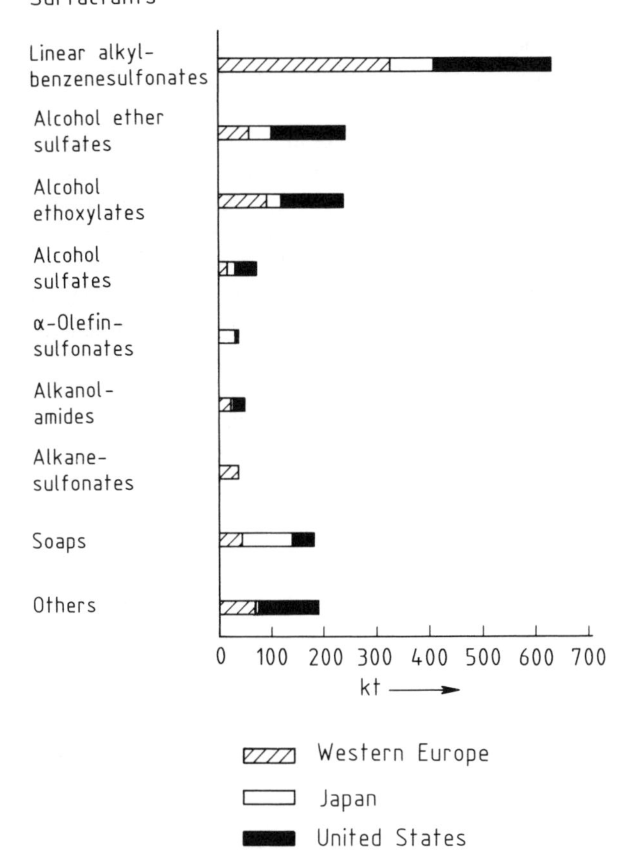

Figure 29. Household surfactant consumption, 1982 [42]

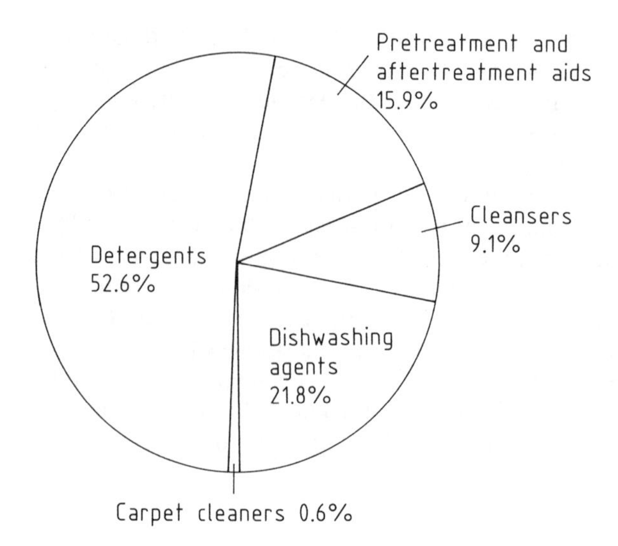

Figure 30. Distribution of surfactant consumption for various types of products in the Federal Republic of Germany [43]

Within the broad category of laundry, dishwashing, and cleansing agents, the quantity of surfactant employed is greatest for the washing of fabrics, as shown in Figure 30 for the Federal Republic of Germany.

Wash technology has been the subject of major changes and developments during the past 30 years. The textile market has also changed, with synthetic fibers playing an increasingly large role. The wide variety of substrates and their differing detergency requirements has forced manufacturers to devise surfactants with a broad spectrum of action. No single surfactant is capable of fulfilling all demands in an optimal way; consequently, the trend has been increasingly toward use of surfactant mixtures, in which the characteristics of each component are intended to supplement those of the others. The widening scope of the demand for surfactants used in detergents not only relates to performance, but also encompasses toxicological, ecological, and economic considerations as well. A surfactant suited for detergent use is currently expected to demonstrate the following characteristics [43], [44]:

specific adsorption
soil removal
low sensitivity to water hardness
dispersion properties
soil antiredeposition capability
high solubility
wetting power
desirable foam characteristics
neutral odor
low intrinsic color
storage stability
favorable handling characteristics

minimal toxicity to humans
favorable environmental behavior
assured raw material supply
economy

3.1.1 Anionic Surfactants

Modern detergents generally contain larger amounts of anionic surfactants than nonionic surfactants. The following discussion is especially concerned with anionic surfactants that are either already widely in use or have favorable characteristics, suggesting their likely presence in various products in the future.

Soap. Soap is no longer as important as it was before the existence of mass-produced synthetic detergents. Although the so-called detergent soaps once contained as much as 40% soap as their sole surfactant, current powdered detergents are formulated with mixtures of far more effective surfactants in considerably smaller proportion. A further reason for the disappearance of soap from cleansing agents is its sensitivity to water hardness, manifested through inactivation due to the formation of lime soap, which tends to accumulate on fabrics and washing machine components. Such accumulation reduces the absorbency of fabrics and their permeability to air, and eventually through "aging" causes laundry to become discolored and to develop a disagreeable odor. The primary function that remains for soap currently is as a foam regulator (cf. Section 3.4.3). Nonetheless, some countries still have soap-based cleansing agents, e.g., South Korea and various African states. In addition, in the United Kingdom, one leading manufacturer still distributes a heavy-duty detergent formulated with soap.

Alkylbenzenesulfonates (ABS). Until the mid-1960s, this largest of the synthetic surfactant classes was most prominently represented by tetrapropylenebenzenesulfonate (TPS):

$$\begin{array}{c} CH_3 \quad\quad CH_3 \quad\quad CH_3\ CH_3 \\ CH{-}CH_2{-}C{-}CH_2{-}CH{-}CH \\ CH_3 \quad\quad C_6H_4 \quad\quad CH_3 \\ SO_3Na \end{array}$$

In the 1950s TPS had largely replaced soap as an active ingredient in detergents. It was later found, however, that the branched side-chain present in TPS prevents the compound from undergoing efficient biodegradation; thus, means were developed to replace it by more biodegradable straight-chain derivatives. Since that time, favorable economic circumstances and good performance characteristics have permitted straight-chain or linear alkylbenzenesulfonate (LAS) to take the lead among laundry detergent surfactants in Western Europe, the United States, and Japan. Nevertheless, a few countries remain in which TPS continues to be used in detergents.

$$H_3C-(CH_2)_n-\underset{\underset{\displaystyle SO_3Na}{|}}{\overset{H}{\overset{|}{C}}}-(CH_2)_m-CH_3 \qquad n+m=7\text{-}10$$

LAS

Apart from its effective performance, LAS has very interesting foaming characteristics, which are of great significance to its use in detergents. Its foaming ability is high, and the foam that is produced is readily stabilized by foam stabilizers, an important factor in the United States and Japanese markets. At the same time, however, LAS can be controlled easily by foam inhibitors, and this is significant with respect to detergents for the Western European market, where drum-type washing machines are common.

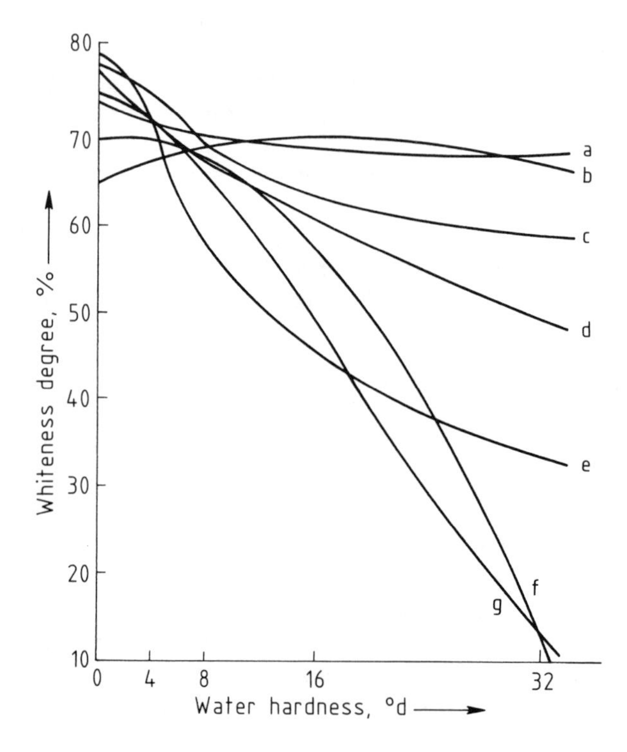

Figure 31. Detergency performance on wool by various surfactants as a function of water hardness [45]
Time: 15 min; temperature: 30 °C; bath ratio: 1:50; concentration: 0.5 g/L surfactant + 1.5 g/L sodium sulfate
a) Nonylphenol 9 EO;
b) C_{12-14} Fatty alcohol 2 EO sulfates; c) C_{15-18} α-Olefinsulfonates; d) C_{16-18} α-Sulfo fatty acid esters; e) C_{12-18} Fatty alcohol sulfates;
f) C_{10-13} Alkylbenzenesulfonates; g) C_{13-18} Alkanesulfonates

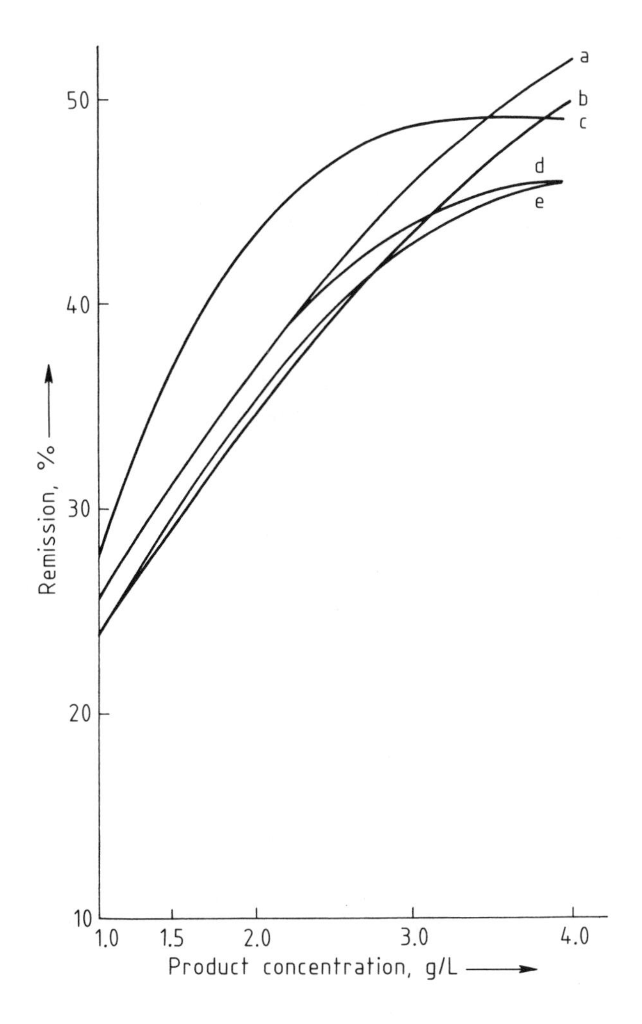

Figure 32. Detergency performance on wool of various anionic surfactants as a function of concentration [43]
Product: 25% surfactant + 75% sodium sulfate; temperature: 30 °C; wash time: 15 min; bath ratio: 1:30; water hardness: 16 °d
a) C_{12-14} Fatty alcohol 2 EO sulfates; b) C_{15-18} α-Olefinsulfonates; c) C_{16-18} α-Sulfo fatty acid esters; d) C_{13-18} Alkanesulfonates; e) C_{10-13} Alkylbenzenesulfonates

As a result of its high solubility, LAS also is employed in formulations for liquid detergents. However, LAS is sensitive to water hardness: the detergency power of LAS diminishes as the hardness of the water increases. The relationship between water hardness and performance for a series of surfactants is well demonstrated by soil removal from wool, as illustrated in Figure 31.

The decline in detergency with increasing water hardness is most dramatic with soap. Sensitivity to water hardness largely disappears in phosphate- and zeolite 4A-containing formulations of the commonly used detergents because of sequestration binding or ion exchange of the hardening agents (cf. Chap. 2, Fig. 24). Figure 32 illustrates wool wash performance in water of average hardness for various readily accessible anionic surfactants in the absence of added complexing agents, plotted as a function of surfactant concentration. It can be seen that those products with a lower sensitivity to hardness display a slight advantage.

Alkanesulfonates (SAS)

$$\underset{\displaystyle SO_3Na}{R^1{-}\underset{|}{C}H{-}R^2} \qquad R^1 + R^2 = C_{11-17}$$

Sodium alkanesulfonates are the compounds that most nearly resemble LAS in detergency properties; therefore, these can largely be substituted for the latter in most formulations.

In contrast to alkyl sulfates, alkanesulfonates are completely insensitive to hydrolysis even at extreme pH values, a result of the presence of a direct carbon–sulfur bond.

The water hardness sensitivity and foaming characteristics of SAS resemble those of LAS as discussed above, apart from slight differences in degree.

α-Olefinsulfonates (AOS)

$$R^1{-}CH_2{-}CH{=}CH{-}(CH_2)_n{-}SO_3Na \qquad \text{Alkenesulfonates}$$

$$R^2{-}CH_2{-}\underset{\displaystyle OH}{\underset{|}{C}H}{-}(CH_2)_m{-}SO_3Na \qquad \text{Hydroxyalkanesulfonates}$$

$$R^1 = C_8{-}C_{12} \qquad n = 1, 2, 3$$

$$R^2 = C_7{-}C_{13} \qquad m = 1, 2, 3$$

Olefinsulfonates are currently prepared commercially starting from α-olefins. Alkaline hydrolysis of the sultone intermediate results in ca. 60–65% alkenesulfonates and ca. 35–40% hydroxyalkanesulfonates. Because of the use of olefinic precursors, these mixtures are customarily called α-olefinsulfonates.

In contrast to LAS and SAS, AOS shows little sensitivity to water hardness (cf. Fig. 31). This apparent advantage is only of significance in a few very special applications, however. Depending on chain length, AOS can cause foaming problems in drum-type washing machines, which requires the addition of special foam inhibitors. Since this problem does not arise with washing machines of the type used in Japan, AOS has long been an important component of Japanese detergents.

In addition to the α-olefinsulfonates, sulfonates prepared from olefins with internal or central double bonds and from vinylidene olefins also exist. These types of olefinsulfonates are unsuitable for detergent use, however, because of their poor performance characteristics. In Figure 33, a C_{15-18} α-olefinsulfonate is compared with a C_{15-18} internal olefinsulfonate, the inferior performance of which is clearly apparent. The reason for the difference is the fact that AOS has its hydrophilic group in a terminal position, whereas with the internal olefinsulfonate, additional isomers can arise in which hydrophilic groups are distributed throughout the entire hydrophobic chain. This has the same consequences as branching (cf. p. 43). It should be noted, however, that olefinsulfonates prepared from internal olefins generally show very good textile wetting characteristics.

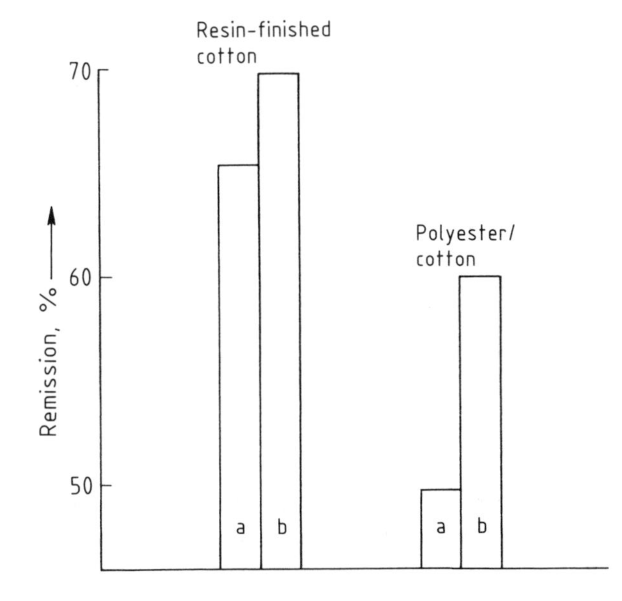

Figure 33. Performance of various olefinsulfonates [46]
Water hardness: 16 °d; wash time: 30 min; wash temperature: 60 °C
a) 0.75 g/L C_{15-18} Internal olefinsulfonates + 2.0 g/L sodium triphosphate;
b) 0.75 g/L C_{15-18} α-Olefinsulfonates + 2.0 g/L sodium triphosphate

α-Sulfo Fatty Acid Esters (SES)

$$R{-}CH(SO_3Na){-}C({=}O){-}OCH_3 \qquad R = C_{14-16}$$

α-Sulfo fatty acid methyl esters

Another important class of anionic surfactants is the α-sulfo fatty acid esters, particularly the methyl derivatives. Apart from their good performance characteristics, α-sulfo fatty acid esters are distinguished by their stability, since the presence of the neighboring sulfonate group reduces any tendency toward hydrolysis of the ester function. Good detergency performance is attained only with products having rather long hydrophobic residues (e.g., stearic acid). The sensitivity of SES to water hardness is small relative to that of LAS and SAS, more nearly resembling that of AOS. One of the interesting detergency properties of α-sulfo fatty acid methylesters is their exceptional dispersion power with respect to lime soap.

Alkyl Sulfates (FAS)

$$R{-}CH_2{-}O{-}SO_3Na \qquad R = C_{11-17}$$

Alkyl sulfates, also known as fatty alcohol sulfates, achieved prominence as early as the 1930s in detergents for *easy care* fabrics (e.g., Fewa) and as components of textile auxiliaries. Their availability resulted from the development by SCHRAUTH of a means for preparing primary fatty alcohols by high-pressure hydrogenation of fatty acids and their methyl esters. Alkyl sulfates possess rather desirable detergency properties, and they are finding increasing application not only in specialty products, but also in heavy-duty detergents.

Alkyl Ether Sulfates (FES)

$$R^2-\underset{|}{\overset{R^1}{C}}H-CH_2-O-(CH_2-CH_2-O)_n-SO_3Na$$

Alkyl ether sulfates

1. $R^1 = H$, $R^2 = C_{10-12}$ — Fatty alcohol ether sulfates

2. $R^1 + R^2 = C_{11-13}$ — Oxo alcohol ether sulfates
   $R^1 = H, C_1, C_2 \ldots$

$n = 1-4$

Alkyl ether sulfates are obtained by ethoxylation and subsequent sulfation of natural and synthetic alcohols. They exhibit the following unique characteristics relative to alkyl sulfates:

low sensitivity to water hardness (cf. Fig. 31)
high solubility
storage stability at low temperature in liquid formulations

Fatty alcohol ether sulfates that are least sensitive to water hardness, e.g., sodium C_{12-14} *n*-alkyl diethylene glycol ether sulfates, actually demonstrate increased detergency performance (e.g., on wool) as the hardness increases. This is a result of the positive electrolyte effects attributable to calcium/magnesium ions. Addition of sodium sulfate produces a slight improvement only in regions of low water hardness. However, detergency performance declines in the presence of sodium triphosphate, which leads to calcium/magnesium sequestration (cf. Fig. 34).

Binding of alkaline-earth ions can occur not only through complexation, but also as a result of ion exchange. Calcium ions can be exchanged particularly effectively and with favorable kinetics through the use of zeolite 4 A. The properties discussed above in the context of calcium sequestration can be equally well described in terms of ion exchange.

From the foregoing, it follows that if one were to undertake the development of, for example, a wool detergent, an important consideration would be the careful choice of supplemental salts; i.e., the presence of complexing agents or ion exchangers is not always advantageous in a detergent formulation containing surfactants if these happen to be insensitive to water hardness. On the other hand, a surfactant that is sensitive to water hardness is not necessarily inferior, provided it is combined with the proper complexing agents or ion exchangers. The situation can be somewhat more complicated, however, if one proposes to develop a detergent for fibers whose nature requires the presence of complexing agents for proper washing.

Alkyl ether sulfates are very intensively foaming compounds that are well suited to use in highly foaming detergents for agitator-type washing machines, but are less

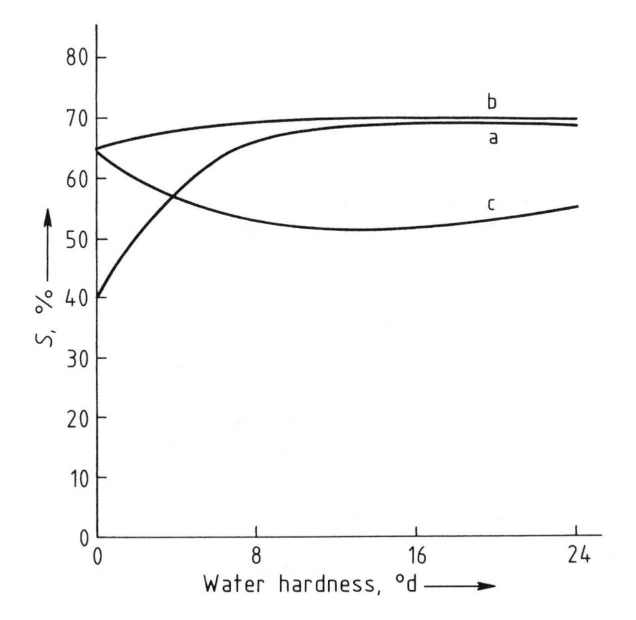

Figure 34. Soil removal S from wool as a function of water hardness at 30 °C with 0.5 g/L sodium C_{12-14} n-alkyl diethylene glycol ether sulfates containing no added electrolyte (a), with 1.5 g/L sodium sulfate (b), and with 1.5 g/L sodium triphosphate (c) [7]

directly applicable to detergents for drum-type machines. Because of their specific properties, alkyl ether sulfates are preferred constituents of easy care and wool detergents, as well as foam baths, hair shampoos, and manual dishwashing agents. The optimal carbon chain length has been established to be C_{12-14} with ca. 2 mol of ethylene oxide.

Analogous to the alkyl sulfates, alkyl ether sulfates have achieved some importance in the United States and Japanese markets. This is because their critical micelle concentration is considerably lower than that for LAS, resulting in very satisfactory washing power even at the low detergent concentrations typical in the United States. In Europe, alkyl ether sulfate use has so far been restricted largely to specialty detergents.

Alkyl ether sulfates were formerly prepared exclusively by ethoxylation and sulfation of natural fatty alcohols. However, synthetic alcohols are also currently employed, particularly the partially branched oxo alcohols and the Ziegler alcohols. The latter are always unbranched due to their mode of synthesis and, thus, show properties largely identical with those of the natural alcohols.

3.1.2 Nonionic Surfactants

Nonionic surfactants of the alkyl polyglycol ether type do not dissociate in aqueous solution, and certain of their properties can be singled out for special attention:

the absence of electrostatic interactions
behavior with respect to electrolytes
the possibility of favorable adjustment of hydrophilic–hydrophobic parameters
anomalous solubility

Adsorption phenomena involving nonionic surfactants can be explained on the basis of hydrophobic interactions, in some cases coupled with steric effects. Electrolytes have no direct influence on adsorption with nonionic surfactants. Nonionic surfactant wash effectiveness is reduced by addition of polyvalent cations, but the cause is reduced negative ζ-potentials of substrate and soil, leading to reduced repulsion and correspondingly poor soil removal [47], [48].

An important advantage of nonionic surfactants that are based on poly(alkylene glycol) ethers as compared to ionic compounds is the fact that a proper relationship can be achieved easily between the hydrophobic and hydrophilic portions of the nonionic surfactants. For example, the hydrophilic portion of the molecule can be extended gradually by stepwise addition of ethylene oxide groups. This leads to stepwise increases in hydration and corresponding successive increases in solubility. On the other hand, with ionic surfactants, the presence of even one ionic group makes such a strong contribution to hydrophilic character that the introduction of further such groups totally eliminates the possibility of any equilibrium relationship with respect to the hydrophobic portion. The result is a rapid decline in detergency properties. For a substance bearing two strong ionic hydrophilic groups to show washing activity, a considerably longer alkyl chain ($> C_{20}$) is generally necessary.

Nonionic surfactants with a given hydrophobic residue can be adjusted to show optimal properties for various substrates with respect to adsorption and wash effectiveness simply by changing the degree of ethoxylation. Wash effectiveness shows an initial increase with an increasing degree of ethoxylation, but a point is then reached after which the wash effectiveness declines markedly (Fig. 35). Textile wetting power often decreases only at very high degrees of ethoxylation, whereas the wetting power for hard hydrophobic surfaces continues to climb as the number of ethoxy groups increases.

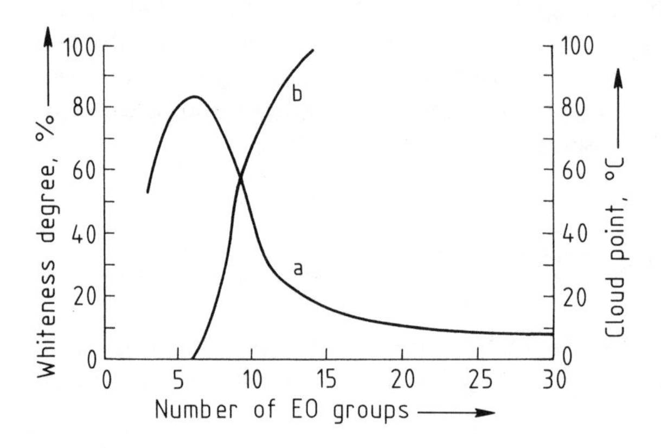

Figure 35. Detergency performance on wool and cloud point of C_{13-15} oxo alcohols as a function of degree of ethoxylation [43]
a) Wool detergency performance (temperature: 30 °C; water hardness: 16 °d; bath ratio: 1:30; time: 15 min; concentration: 0.75 g/L surfactant +2.25 g/L sodium sulfate);
b) Cloud point (surfactant concentration 10 g/L)

Table 13. Detergency performance (in % remission) of alcohol ethoxylates with comparable cloud points [45]

Surfactant*	Cloud point,	Cotton		Resin-finished cotton		Polyester/ cotton	
	°C	60 °C	90 °C	60 °C	90 °C	60 °C	90 °C
C_{11-15} *sec*-Alcohol 9 EO	59	71	53	69	55	65	48
C_{9-11} Oxo alcohol 7 EO	61	57	44	66	52	53	43
Oleyl/cetyl alcohol 10 EO**	89	60	68	70	72	57	67
C_{13-15} Oxo alcohol 11 EO	88	54	69	67	69	53	58

* Surfactant concentration: 0.75 g/L; water hardness: 16 °d; wash time: 30 min.
** Iodine number 45.

Nonionic surfactants based on poly(alkylene glycol) ether show a solubility anomaly: when they are heated in aqueous solution, turbidity appears, usually at a relatively precise temperature. The result is a separation into two phases, one with a high water content and one with a low water content. The corresponding characteristic temperature for a given surfactant is known as its cloud point. The cloud point moves to a higher temperature as the number of ethoxy groups increases. If the cloud point is not greatly exceeded, then the largely aqueous phases and largely surfactant phases form an emulsion. For a given surfactant, adsorption (thus, washing power) decreases when the cloud point is surpassed to a significant extent. The main reason for this behavior is the diminished solubility of the washing active homologues, which are then expelled from the aqueous phase. Nevertheless, nonionic surfactants with a cloud point somewhat below the application temperature commonly show better performance than those whose cloud point is higher (Fig. 35). Thus, the application temperature is a significant factor in determining an optimal degree of ethoxylation (Table 13). The table shows that the best detergency performance is obtained when the temperature is maintained near the cloud point.

The cloud point can be greatly reduced by addition of several grams of electrolytes per liter, depending on the surfactant. It must be emphasized, however, that everything said in this section is strictly applicable only to systems comprised of pure nonionic surfactants. For binary mixtures of nonionic and ionic surfactants, it is important to recognize that even a small amount of ionic surfactant can cause the cloud point to rise more or less dramatically.

The share of nonionic surfactants in overall surfactant production has been increasing for a number of years. The major contributors to this increase have been fatty alcohol, oxo alcohol, secondary alcohol, and alkylphenol ethoxylates, all of which are obtained by reaction of the corresponding hydroxy compounds with ethylene oxide.

The reasons for the increased use of nonionic surfactants are found in their favorable detergency properties, particularly with respect to synthetics and at the

Table 14. Critical micelle concentration c_M of various surfactants [45]

Surfactant	c_M, g/L
LAS (C_{10-13} alkyl)	0.65
C_{12-17} Alkanesulfonates	0.35
C_{15-18} α-Olefinsulfonates	0.30
C_{12-14} Fatty alcohol 2 EO sulfates	0.30
Nonylphenol 9 EO	0.049
Oleyl/cetyl alcohol 10 EO*	0.035

* Iodine number 45.

decreasing wash temperatures dictated by energy conservation measures in Western Europe and the United States.

Favorable detergency properties of nonionic surfactants derive largely from the following factors:

low critical micelle concentration (c_M)
very good detergency performance
soil antiredeposition characteristics with synthetic fibers

The low c_M values of nonionic surfactants mean that they display high detergency performance even at relatively low concentrations. Table 14 provides data illustrating the low c_M of these compounds relative to that of anionic surfactants.

Alkylphenol Polyglycol Ethers (APEO). Alkylphenol polyglycol ethers based on octyl-, nonyl-, and dodecylphenol poly(ethylene glycol) ethers achieved early success due to their exceptional detergency properties, particularly their oil and fat removal characteristics.

$$R-C_6H_4-O-(CH_2-CH_2-O)_nH \qquad R = C_{8-12} \quad n = 5-10$$

The significance of APEOs has greatly declined, however, especially in the Federal Republic of Germany, due to a debate over their environmental characteristics, particularly the extent of their biodegradability and the fish toxicity of certain metabolites resulting from partial biodegradation.

Alkyl Polyglycol Ethers (AEO). Alkyl polyglycol ethers based on natural and synthetic (e.g., oxo or Ziegler) alcohols have become standard components of modern detergents, present to greater or lesser extent in practically all detergent formulations.

$$\begin{array}{l} \quad R^1 \\ \quad | \\ R^2-CH-CH_2-O-(CH_2-CH_2-O)_nH \end{array}$$

$R^1 = H \quad R^2 = C_{6-16} \quad n = 3-15$

Fatty alcohol polyglycol ethers

$R^1 + R^2 = C_{7-13} \quad R^1 = H, C_1, C_2 \ldots \quad n = 3-15$

Oxo alcohol polyglycol ethers

One or more ethylene oxide groups can also be replaced by propylene oxide residues. Such a change can increase the hydrophobic character of the alkyl group and permit modification of foam characteristics.

Fatty Acid Alkanolamides (FAA). Important fatty acid alkanolamides are ethanolamides of fatty acids with the following structure:

$$R-\overset{O}{\overset{\|}{C}}-N\begin{cases}(CH_2-CH_2-O)_nH \\ (CH_2-CH_2-O)_mH\end{cases} \qquad \begin{array}{l} R = C_{11-17} \\ n = 1, 2 \\ m = 0, 1 \end{array}$$

Their most important function is to act as foam boosters, adding desired stability to the foam produced by detergents prone to heavy foaming. This property is no longer desirable for the drum-type washing machines employed in Western Europe. Nevertheless, compounds in this class have taken on new meaning even in Western Europe, where they are employed as detergency boosters, particularly in low-temperature applications. A small amount of such material is capable of enhancing the soil removal properties of the classical detergent components.

Amine Oxides. The amine oxides produced by oxidation of tertiary amines with hydrogen peroxide are compounds that exhibit cationic behavior in acidic conditions ($pH < 3$), but they behave as nonionic surfactants under neutral or alkaline conditions. For this reason they are included in the nonionic surfactant category.

$$R-\overset{CH_3}{\overset{|}{\underset{CH_3}{\underset{|}{N}}}}\rightarrow O \qquad R = C_{12-18}$$

Compounds in this class have been known since 1934 and were described as detergent components in a patent issued to IG-Farbenfabriken [49]. Combinations of alkylbenzenesulfonates and specific amine oxides are reputed to be especially gentle to the skin. Despite good detergency properties, however, the compounds are found almost exclusively in specialty detergents. The reasons for this include high cost, low thermal stability, and high foam stability.

3.1.3 Cationic Surfactants

Long-chain cationic surfactants such as distearyldimethylammonium chloride (DSDMAC) exhibit extraordinarily high sorption power with respect to a wide variety of surfaces [50]–[53]. Figure 36 shows this behavior in the case of several textile fibers. Adsorption rises steeply at low surfactant concentrations, followed by rapid saturation as the concentration increases. This behavior suggests complete coverage of boundary surfaces.

At the same time, cationic surfactants display behavior opposite that of anionic surfactants as regards charge relationships on solids. Since the surfactant molecules bear a positive charge, their adsorption reduces the negative ζ-potential of solids present in aqueous solution, thereby reducing mutual repulsions, including that between soil and fibers. Use of higher surfactant concentrations causes charge reversal; thus, solid particles become positively charged, resulting again in repulsion. Soil removal can be achieved, if adequate amounts of cationic surfactants are present and if their alkyl chains are somewhat longer than those of comparable anionic surfactants. This fact has little practical significance, however, since the subsequent rinse

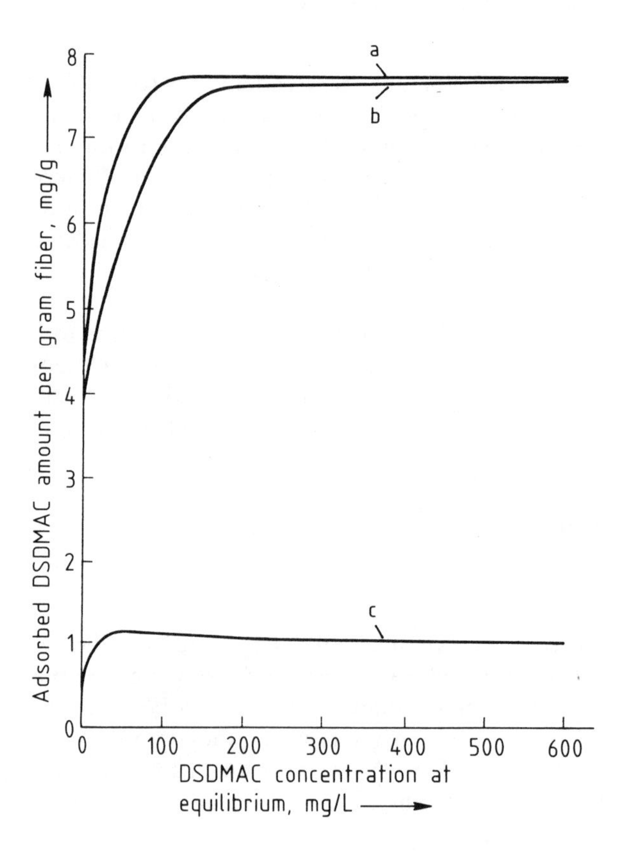

Figure 36. Adsorption isotherms of distearyldimethylammonium chloride (DSDMAC) on wool (a), cotton (b), and polyacrylonitrile (c) [50] Temperature: 23 °C; time: 20 min; bath ratio: 1:10

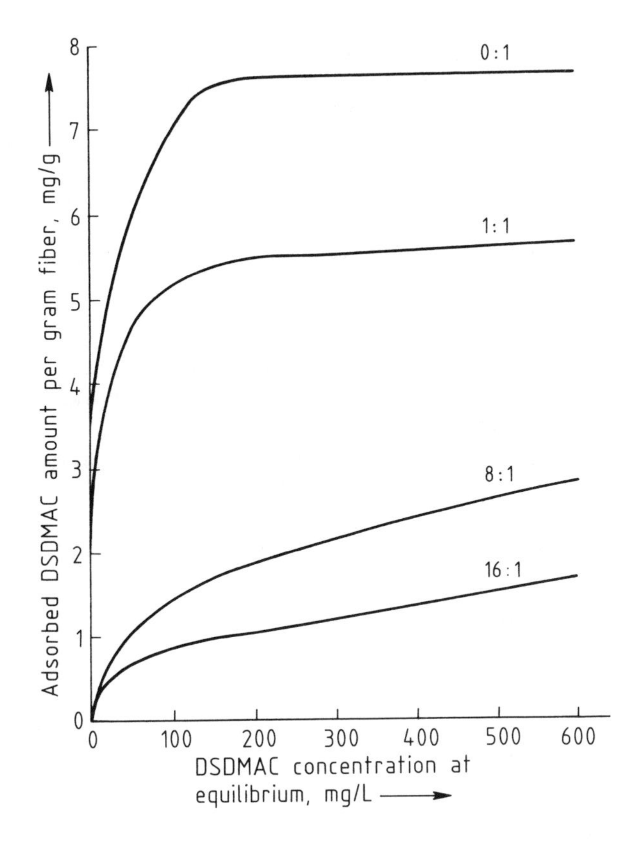

Figure 37. Adsorption isotherms of distearyldimethylammonium chloride (DSDMAC) on cotton as a function of the alkyl polyglycol ether:DSDMAC ratio at equilibrium [50]
Temperature: 23 °C; time: 20 min; bath ratio: 1:10

and dilution processes cause charge reversal in the direction of negative ζ-potentials, whereby a large fraction of the previously removed soil is once more attracted to the fibers. Therefore, cationic surfactants are employed in laundry and cleansing agents only for the purpose of achieving certain special effects, which include applications in fabric softeners, antistatic agents, and microbicides.

Mixtures made up of equivalent amounts of anionic surfactant and cationic surfactant remain virtually unadsorbed on surfaces and thus display no washing effect. Reactions between anionic and cationic surfactants produce neutral salts with extremely low water solubility. Regarding washing, these behave like an additional burden of greasy soil. On the other hand, addition of small amounts of certain specific cationic surfactants to an anionic surfactant — or even a nonionic surfactant — can enhance detergency performance.

Nonionic surfactants are more tolerant of the presence of cationic surfactants. Mixtures of the two are sometimes used in specialty detergents intended to have a fabric softening effect. In such cases, one must take into account the fact that adsorption of the cationic surfactant can be greatly reduced by the presence of the nonionic surfactant, depending on the concentration of the latter, a phenomenon that can have a negative influence on the fabric softening characteristics (Fig. 37).

Dialkyldimethylammonium Chlorides. The first surfactant developed in this category was distearyldimethylammonium chloride (DSDMAC), introduced in 1949 as a fabric softener for diapers and presented to the United States market a year later in its present form as a laundry rinse fabric softener.

$$\left[\begin{array}{ccc} R & & CH_3 \\ & \overset{+}{N} & \\ R & & CH_3 \end{array}\right] Cl^- \qquad R = C_{16-18}$$

Only in the mid-1960s did these surfactants begin to have a major impact as laundry aftertreatment aids on the United States and Western European markets (especially in the Federal Republic of Germany).

A recent application for DSDMAC, albeit a minor one, is as a wash cycle fabric softener in specialty detergents.

Imidazolinium Salts. Imidazolinium salts of the type 1-(alkylamidoethyl)-2-alkyl-3-methylimidazolinium methyl sulfate have achieved a place as rinse softening agents, although not nearly as significant as that of DSDMAC.

$$\left[\begin{array}{l} \quad\;\; CH_3 \\ \quad\;\; | \\ \quad\;\; N-CH_2 \\ R-C^{+} \quad | \\ \quad\;\; N-CH_2 \qquad H \;\; O \\ \quad\;\; | \qquad\qquad\quad\; | \;\;\; \| \\ \quad\;\; CH_2-CH_2-N-C-R \end{array}\right] CH_3O-SO_3^- \qquad R = C_{16-18}$$

Alkyldimethylbenzylammonium Chlorides. Compounds of the alkyldimethylbenzylammonium chloride type show only limited fabric softening character, but they are used in laundry disinfecting agents as a result of their activity toward gram-positive and gram-negative bacteria.

$$\left[\begin{array}{c} CH_3 \\ | \\ R-\overset{+}{N}-CH_2-C_6H_5 \\ | \\ CH_3 \end{array}\right] Cl^- \qquad R = C_{8-18}$$

The high adsorption capability of alkyldimethylbenzylammonium chloride also leads to applications as antistatic agents in laundry aftertreatment products.

3.1.4 Amphoteric Surfactants

Compounds of the alkylbetaine or alkylsulfobetaine type possess both anionic and cationic groups in the same molecule even in aqueous solution. Despite what could be seen in some respects as excellent detergency properties, these substances are only rarely employed in specialty detergents, the reasons being primarily economic.

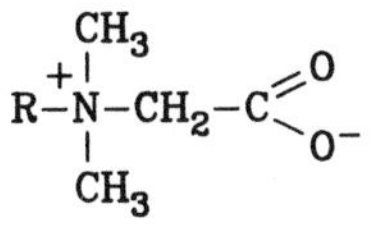

$$R-\overset{+}{N}(CH_3)_2-CH_2-C\begin{matrix}\nearrow O \\ \searrow O^-\end{matrix} \qquad R = C_{12-18} \quad \text{Alkylbetaines}$$

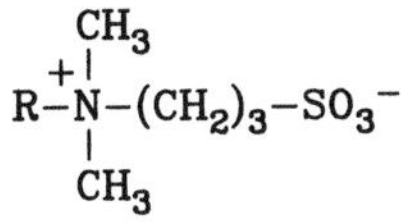

$$R-\overset{+}{N}(CH_3)_2-(CH_2)_3-SO_3^- \qquad R = C_{12-18} \quad \text{Alkylsulfobetaines}$$

3.2 Builders

Detergent builders play a central role in the course of the washing process [54], [55]. Their function is largely that of supporting detergent action and of eliminating calcium and magnesium ions, which arise partly from the water and sometimes also from soil and fabrics.

The category of builders is comprised of several types of materials: specific alkaline substances such as sodium carbonate and sodium silicate; complexing agents like sodium diphosphate–sodium triphosphate or nitrilotriacetic acid (NTA); and ion exchangers, such as water-soluble polycarboxylic acids and insoluble zeolites (e.g., zeolite 4 A).

Modern detergent builders must fulfill a number of criteria [7]:

1) Elimination of alkaline-earth ions from
   - water
   - textiles
   - soil

2) Single wash cycle performance
   - high specific detergency for pigments and fats
   - distinct detergency on specific textile fibers
   - enhancement of surfactant properties
   - dispersion of soil in detergent solutions
   - favorable influence on foam characteristics

3) Multiple wash cycle performance
   good soil antiredeposition capability
   prevention of incrustations on textiles
   prevention of deposits in washing machines
   favorable anticorrosion properties

4) Commercial properties
   chemical stability
   industrial handleability
   no hygroscopic tendencies
   optimal color and odor qualities
   compatibility with other detergent ingredients
   storage stability
   assured raw material basis

5) Human toxicological safety assurance

6) Environmental properties
   response to deactivation by biological degradation, adsorption, or other mechanisms
   no negative influence on the biological systems found in sewage plants and surface water
   no uncontrolled accumulation
   no heavy-metal remobilization
   no eutrophication
   no detrimental effects on drinking water quality

7) Economy

3.2.1 Alkalies

Alkalies such as potash and soda have been used to enhance the washing effectiveness of water since antiquity. Their activity is based on the fact that soil and fibers become more negatively charged as the pH increases, resulting in increased mutual repulsion. Alkali also precipitates ions that contribute to water hardness.

At the beginning of the 20th century, the principal ingredients (apart from soap) of all detergents were soda and silicate, which often comprised nearly 50% of the formulation of a powdered detergent. These substances were somewhat replaced during the 1930s by sodium monophosphate, and currently, at least in Western Europe, a household detergent that contained only soda or monophosphate as builders would be classed as outmoded. Modern builders no longer precipitate water-hardening agents; instead, the ions that make water hard are removed by complexation (sequestration) or ion exchange. Even commercial laundries using soft water have converted to detergents that are low in soda and contain the complexing

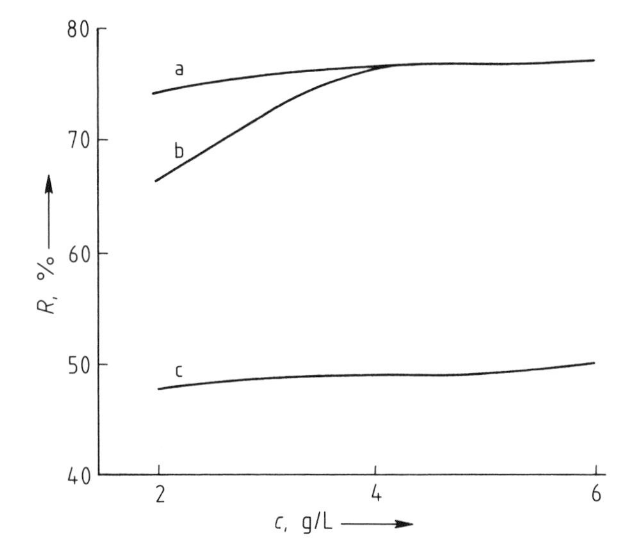

Figure 38. Comparison of soil removal (given as remission *R* in %) on resin-finished cotton at 0 °d and 90 °C for sodium triphosphate- and sodium carbonate-containing detergents, formulated on the basis of
a) 40% Sodium triphosphate;
b) 20% Sodium triphosphate + 20% sodium sulfate;
c) 40% Sodium carbonate

agent sodium triphosphate. The latter has many valuable properties, among which are better detergency performance than soda or sodium monophosphate (Fig. 38) [56].

3.2.2 Complexing Agents

Soda and sodium monophosphate induce precipitation of calcium and magnesium salts from tap water. This can lead to troublesome depositions on both laundry and laundering apparatus. By contrast, sequestering agents form somewhat stable, water-soluble complexes with alkaline-earth elements, as well as with the traces of heavy metals present in water. Often the resulting complexes are chelates (Fig. 39) [57]–[61].

Temperature and complexing agent concentration are generally the decisive factors in successful elimination of polyvalent metal ions. Table 15 shows the calcium binding capacity of various sequestering agents as a function of temperature. Calcium binding capacity is here to be viewed as a quantitative measure of the stoichiometry of the resulting complexes. For most sequestering agents, this capacity decreases markedly with increasing temperature.

Regarding stability, the data presented show only that the stability constants (which are a function of the method of study employed, e.g., the dissolving power for freshly precipitated calcium carbonate) exceed a specific value that is dependent on the solubility of the calcium carbonate. Extremely high stability constants or stabilities, which would result in very low calcium ion concentrations, are not re-

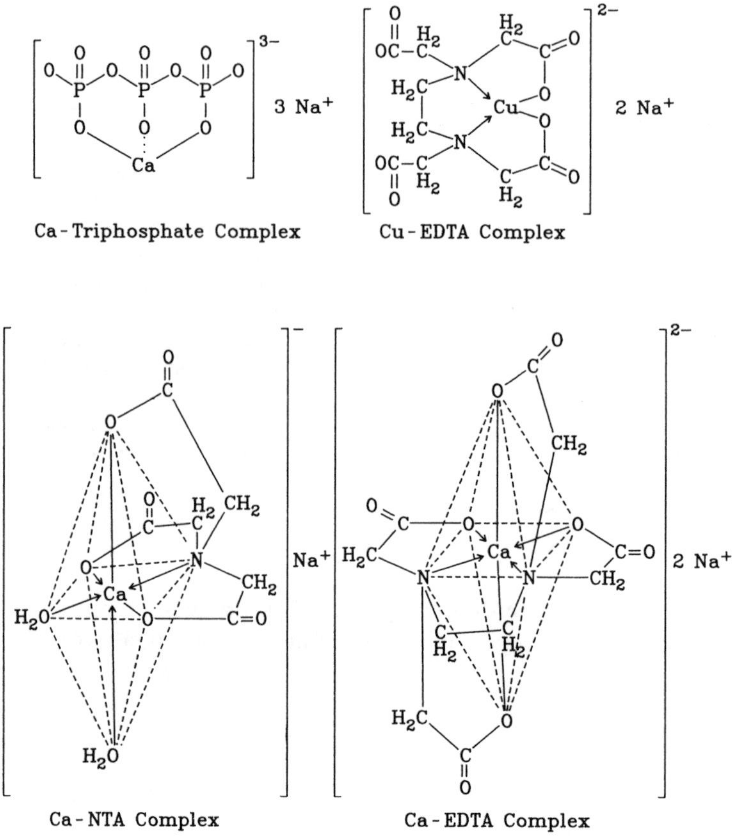

Ca-Triphosphate Complex

Cu-EDTA Complex

Ca-NTA Complex

Ca-EDTA Complex

Figure 39. Metal complexes

quired; indeed, they are normally undesirable. It is important that those salts present to the greatest extent, such as calcium carbonate or others with even higher solubility products, be prevented from precipitating during the washing process. Less soluble calcium salts normally play a minor role.

Relative to other polyvalent ions, alkaline-earth ion sequestration is of primary concern because these ions are most likely to be present in high concentration. Nevertheless, heavy-metal ions must also be eliminated because even in trace amounts their presence can have a negative effect on the washing process. For this reason, low concentrations of selective complexing agents are generally added as well, usually EDTA (ethylenediaminetetraacetic acid) or specific phosphonic acids, e.g., nitrilotrimethylenephosphonic acid.

If a sequestering agent is present in less than stoichiometric amounts relative to polyvalent metal ions, precipitation of carbonates and insoluble salts of the sequestering agent with the water hardeners normally results. Even with adequate amounts of sequestering agent, this effect is important because dilution during the rinse cycle

Table 15. Calcium binding capacity of selected sequestering agents [7]

Structure	Chemical name	Calcium binding capacity, mg CaO/g	
		20 °C	90 °C
$NaO-P(=O)(ONa)-O-P(=O)(ONa)-ONa$	sodium diphosphate	114	28
$NaO-P(=O)(ONa)-O-P(=O)(ONa)-O-P(=O)(ONa)-ONa$	sodium triphosphate	158	113
$HO-P(=O)(OH)-C(OH)(CH_3)-P(=O)(OH)-OH$	1-hydroxyethane-1,1-diphosphonic acid	394	378
$N(CH_2-PO_3H_2)_3$	nitrilotrimethylenephosphonic acid	224	224
$N(CH_2-C(=O)-OH)_3$	nitrilotriacetic acid	285	202
$N(CH_2-C(=O)-OH)_2(CH_2-CH_2-OH)$	*N*-(2-hydroxyethyl)iminodiacetic acid	145	91

Table 15. (continued)

Structure	Chemical name	Calcium binding capacity, mg CaO/g	
		20 °C	90 °C
$HO-C(=O)-H_2C$, $HO-C(=O)-H_2C$ $>N-(CH_2)_2-N<$ $CH_2-C(=O)-OH$, $CH_2-C(=O)-OH$	ethylenediaminetetraacetic acid	219	154
cyclopentane ring bearing four $-C(=O)-OH$ groups (HO–C, C–OH), with H atoms	1,2,3,4-cyclopentanetetra-carboxylic acid	280	235
$HO(O=)C-H_2C-C(OH)(C(=O)OH)-CH_2-C(=O)OH$	citric acid	195	30
$(HO-C(=O))_2CH-O-CH_2-C(=O)-OH$	*O*-carboxymethyltartronic acid	247	123
$HO-C(=O)-CH_2-CH(-O-CH_2-C(=O)-OH)-C(=O)-OH$	*O*-carboxymethyloxysuccinic acid	368	54

can cause sequestrant concentration to fall below the necessary value, thereby permitting undesirable precipitates to form. These can build up on both fabric and washing machine components, leading eventually to serious accumulations. The problem can be particularly severe if conditions permit large crystals to form as a result of seed crystals located on fabric or machine components.

There are a number of compounds that even in substoichiometric amounts are capable of retarding, hindering, or otherwise interfering with precipitation of insoluble salts. In some cases their action induces salts to precipitate in amorphous form, thereby reducing the tendency toward formation of crystals such as calcite, whose sharp edges can be damaging to fabrics. Sodium triphosphate, the dominant additive in modern detergents, strongly exhibits the latter property even at low concentrations; this is known as a *threshold effect*.

Despite the many desirable properties shown by sodium triphosphate in the washing process, its continued use has been the subject of an international debate within the industry for many years. The problem is that sodium triphosphate is a contributor to eutrophication of standing or slowly flowing surface waters; that is, it may lead to overfertilization, which in turn encourages extreme algal growth and adversely influences marine organisms [62], [63]. Recognition of the problem has led to an intense worldwide search for suitable replacements [64]–[77]. Developments have been concentrated not only on sequestering agents, but also on ion exchangers, since these are capable of binding polyvalent metal ions [64], [66], [68], [69], [72], [78]–[82]. Most of the promising substitutes examined so far are organic compounds, primarily those derived from raw materials produced by the petrochemical industry. Few of these substances are available in large quantity. The cost for commercial production on the necessary scale of $>10^6$ t/a would in many cases be prohibitive. Crude cost estimates suggest that production cost for organic compounds in general significantly exceeds those of the common inorganic sequestering agents.

Apart from eliminating troublesome cations and achieving good single wash cycle performance, other important factors in the washing process are dispersion of soil and prevention of soil redeposition (cf. Chap. 2). Because of the presence of distinct localized charges, sequestering agents are readily adsorbed onto pigmented soil. As a result, these compounds often act as effective dispersing agents for such soils.

Figures 40 and 41 show the relationship between dispersing agent concentration and sediment volume for two structurally different soil pigments where the sediment volume is understood as an indicator of the effectiveness of the dispersion process [83]. Voluminous sediments are formed in the presence of poor dispersing agents; adsorption is insufficient to produce an adequate negative surface charge, and the result is coagulation. Figure 40 shows that the extent of dispersion, in this case of kaolin, depends on the specificity with which a sequestering agent is adsorbed. Virtually no effect is produced by sodium sulfate (a) and a nonionic surfactant (b). Anionic surfactants begin to show stabilization only at high concentrations that are rarely used in practice (f).

With graphite, the results are exactly opposite those observed with kaolin (Fig. 41). Sodium triphosphate alone shows no effect and is, thus, analogous to sodium sulfate (curves a and b), whereas sodium *n*-dodecyl sulfate has a very significant stabilizing effect. Because *n*-dodecyl sulfate is sensitive to water hardness, the dispersing effect increases with diminishing hardness; hence, the sediment volume decreases (both curves c: 16 and 0 °d). Although neither electrolytes nor sequestering agents are alone capable of stabilizing the pigment soil graphite, they are capable of

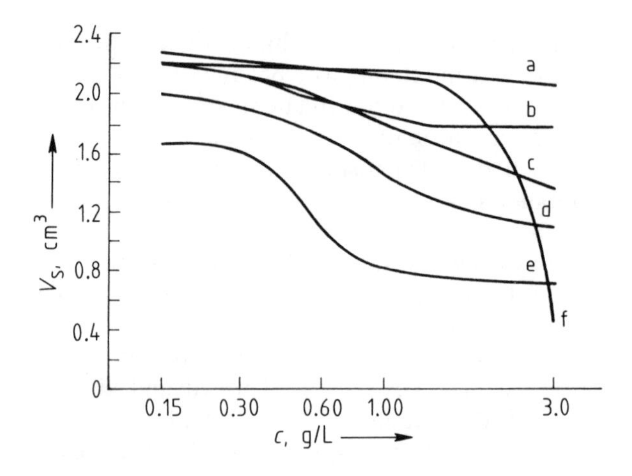

Figure 40. Sediment volume V_s of kaolin as a function of active ingredient concentration for water of hardness 16 °d [7]
a) Sodium sulfate; b) C_{16-18} *n*-Alkyl decaglycol ethers; c) Hydroxyethyliminodiacetic acid, sodium salt; d) Nitrilotriacetic acid, sodium salt; e) Sodium triphosphate; f) Sodium *n*-dodecyl sulfate

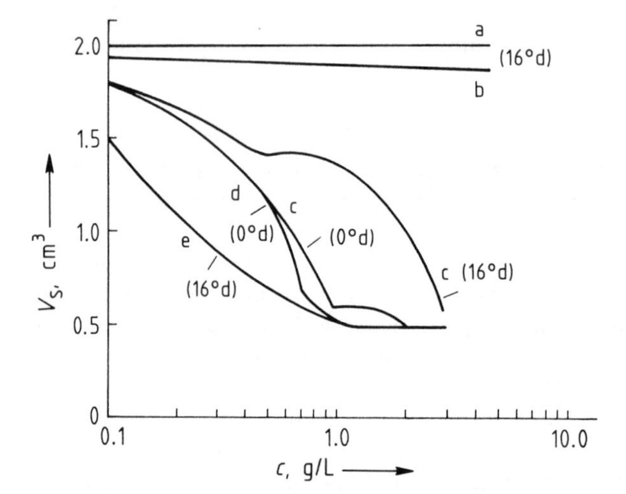

Figure 41. Sediment volume V_s of graphite as a function of active ingredient concentration [7]
a) Sodium sulfate; b) Sodium triphosphate; c) Sodium *n*-dodecyl sulfate; d) Sodium *n*-dodecyl sulfate + 1.5 g/L sodium sulfate; e) Sodium *n*-dodecyl sulfate + 1.5 g/L sodium triphosphate

exerting a positive influence, either indirectly through an electrolyte effect (curve d) or in hard water as a result of fixation of alkaline-earth ions (curve e). With sodium triphosphate, sequestration is accompanied not only by an electrolyte effect, but also a pH effect. An increase in the hydroxide ion concentration and accompanying hydroxide ion adsorption leads to increased electrostatic repulsion and, thus, to better dispersion (reduced sediment volume). The indirect effect of sequestering agents on the dispersion of hydrophobic pigments applies also in principle to emulsification of water-insoluble greasy soil.

A number of additional factors must be considered in selecting a sequestering agent for detergent use. Several important criteria have been listed on p. 63. The extent to which evaluation of a sequestering agent can vary depends on how the several criteria are weighted, as is apparent from Table 16.

Table 16. Evaluation of sequestering agents according to various criteria

Substance (as sodium salt)	Calcium sequestration	Single wash cycle performance	Incrustation on fabrics and washing machines	Hygroscopicity
Sodium diphosphate	low	very good	very heavy	insignificant
Sodium triphosphate	adequate	very good	little	insignificant
1-Hydroxyethane-1,1-diphosphonic acid	very high	very good	very little	hygroscopic
Nitrilotrimethylenephosphonic acid	high	good	little	very hygroscopic
Nitrilotriacetic acid	high	very good	very little	hygroscopic
N-(2-hydroxyethyl)iminodiacetic acid	less adequate	good		very hygroscopic
Ethylenediaminetetraacetic acid	high	good	very little	very hygroscopic
1,2,3,4-Cyclopentanetetracarboxylic acid	high	fair	little	hygroscopic
Citric acid	low	fairly low	little	insignificant
O-Carboxymethyltartronic acid	high	good	little	hygroscopic
O-Carboxymethyloxysuccinic acid	less adequate	fair		hygroscopic

3.2.3 Ion Exchangers

As has already been frequently observed, the disruptive effect of polyvalent metal ions can be reduced not only by the use of low molecular mass sequestering agents, but also by ion exchangers [64], [66], [69], [72], [78]–[82]. Table 17 shows that ion exchangers generally have a high binding capacity for calcium but that this usually decreases with increasing temperature.

Until a few years ago the idea of introducing water-insoluble substances into detergents had never been seriously investigated. This was primarily because the known materials lacked sufficient calcium binding ability, were available only in forms with unsuitable particle structure, and were deemed impractical for economic reasons. Success was first achieved through research in the field of sodium aluminum silicates. Systematic investigation revealed that, among the many known types of sodium aluminum silicates, those with a regular crystalline form were appropriate for use in the washing process. One particular modification proved especially applicable and economically interesting: zeolite 4 A [35]–[38], [85]–[91], occasionally referred to in the literature as HAB — from the German "heterogener anorganischer builder" (heterogeneous inorganic builder) — and sold under the trade name Sasil (sodium **a**luminum **sil**icate) [81], [82]. The ion-exchange behavior of this particular sodium aluminum silicate (Table 17) depends on ionic size and on the state of hydration of the ions. In addition to Ca and Mg, exchange also takes place with Pb,

Table 17. Calcium binding capacity of selected ion exchangers [84]

Formula	Chemical name	Water solubility*	Calcium binding capacity**, mg CaO/g	
			20 °C	90 °C
$[-CH_2-CH(COOH)-]_n$	Poly(acrylic acid)	s	310	260
$[(-CH_2-CH(COOH)-)_x\ (-CH_2-CH(CH_2-OH)-)_y]_n$	Poly(acrylic acid-*co*-allyl alcohol)	s	250	140
$[(-CH_2-CH(COOH)-)_x\ (-CH(HOOC)-CH(COOH)-)_y]_n$	Poly(acrylic acid-*co*-maleic acid)	s	330	260
$[-CH_2-C(OH)(COOH)-]_n$	Poly(α-hydroxyacrylic acid)	s	300	240
$[-CH_2-CH_2-CH(HOOC)-CH(COOH)-]_n$	Poly(tetramethylene-1,2-dicarboxylic acid)	s	240	240
$[-CH(OCH_3)-CH_2-CH(HOOC)-CH(COOH)-]_n$	Poly(4-methoxytetramethylene1,2-dicarboxylic acid)	s	430	330
$x\,Na_2O \cdot Al_2O_3 \cdot y\,SiO_2 \cdot z\,H_2O$	Sodium aluminum silicate	i	165	190

* s = soluble, i = insoluble. ** Sodium salts.

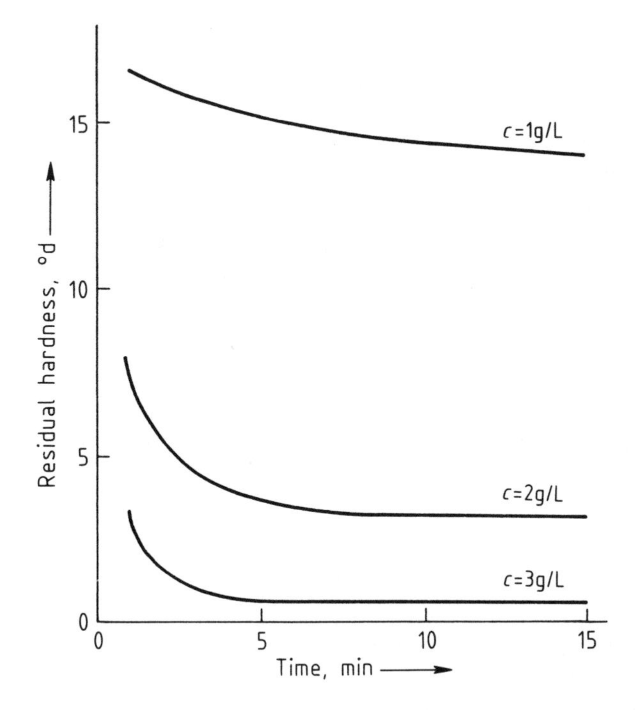

Figure 42. Kinetics of calcium binding to a sodium aluminum silicate (zeolite 4 A) [92]
Initial hardness: 30 °d; temperature: 25 °C

Cu, Ag, Cd, Zn, and Hg ions. The elimination of calcium ions — and to a lesser extent magnesium ions — is of greatest importance for the washing process, but ion exchange capacity for heavy metal ions is also important from an ecological standpoint [89].

Ion exchange is dependent not only on ionic size, but also on concentration, time, temperature, and pH. Figure 42 shows the exchange kinetics as a function of concentration and time. Calcium ions are exchanged very rapidly. The process occurs somewhat more slowly with magnesium, although exchange becomes more rapid at higher temperature. These results can be explained by the larger hydration shell of magnesium ions that hinders exchange at low temperature but is destroyed at higher temperature.

The fact that ion exchange occurs in a heterogeneous phase and that adsorption–desorption phenomena are absent means it is advantageous to employ combinations of sodium aluminum silicate with water-soluble sequestering agents. The latter have the ability to remove polyvalent cations, especially calcium and magnesium, from a solid surface, transport them through an aqueous medium, and then release them to the ion exchanger. This property is best described as a *carrier effect* consisting of the following steps (cf. also Fig. 21, p. 31):

1) Sorption of the carrier onto the fiber–soil boundary surface
2) Complexation of calcium and magnesium ions
3) Transport from the solid fiber–soil boundary surface through the wash liquor to the sodium aluminum silicate boundary surface

4) Dissociation of the complex into carrier and calcium and magnesium ions
5) Ion exchange of calcium and magnesium ions for sodium ions in the sodium aluminum silicate crystal
6) Renewed sorption of the carrier on the fiber–soil boundary surface and further complexation with calcium and magnesium ions

If the properties of sodium triphosphate (a water-soluble sequestering agent) are compared with those of sodium aluminum silicate (a water-insoluble ion exchanger), a number of similarities become apparent, as do typical differences [92]:

1) *Properties of sodium triphosphate:*
   complex formation with polyvalent ions
   alkaline reaction
   specific adsorption on pigments and fibers
   specific electrostatic charging of pigments and fibers, dissolution of polyvalent ions from soil and fibers
   nonspecific electrolyte effect

2) *Properties of sodium aluminum silicate:*
   binding capability for polyvalent ions through ion exchange
   alkaline reaction
   adsorption of molecularly dispersed substances
   heterocoagulation with pigments
   crystallization surface for poorly soluble compounds

The different properties of sodium triphosphate and sodium aluminum silicate can be explained on the basis of their differing solubilities and the differences in the ways they eliminate ions contributing to water hardness. In the case of sodium triphosphate, calcium binding occurs by chelation, whereas with sodium aluminum silicate, calcium binding is a result of ion exchange [92].

Ca^{2+}-binding through chelation (sodium triphosphate):

```
 ┌   O       O       O   ┐5−                ┌   O       O       O   ┐3−
 │O  ‖  O    ‖  O    ‖  O│                   │O  ‖  O    ‖  O    ‖  O│
 │ \ P /  \  P /  \  P / │  5 Na+   Ca2+     │ \ P /  \  P /  \  P / │  3 Na+
 │   |       |       |   │        ───────>   │   |       |       |   │
 └   O       O       O   ┘                   │   O       O       O   │
                                             │     \     :     /     │
                                             └         Ca            ┘
```

Ca^{2+}-binding through ion exchange (zeolite 4A):

```
┌     |     |      ┐            ┌     |     |      ┐
│     O     O      │            │     O     O      │
│     |     |      │            │     |     |      │
│ −O−Si−O−Al⁻−O−   │    Ca2+    │ −O−Si−O−Al⁻−O−   │
│     |     |      │  ──────>   │     |     |      │   + 2 Na+
│     O 2Na+ O     │            │     O Ca2+ O     │
│     |     |      │            │     |     |      │
│ −O−Al⁻−O−Si−O−   │            │ −O−Al⁻−O−Si−O−   │
│     |     |      │            │     |     |      │
│     O     O      │            │     O     O      │
└     |     |      ┘            └     |     |      ┘
```

Figure 43. Scanning electron micrograph of zeolite 4A [87]

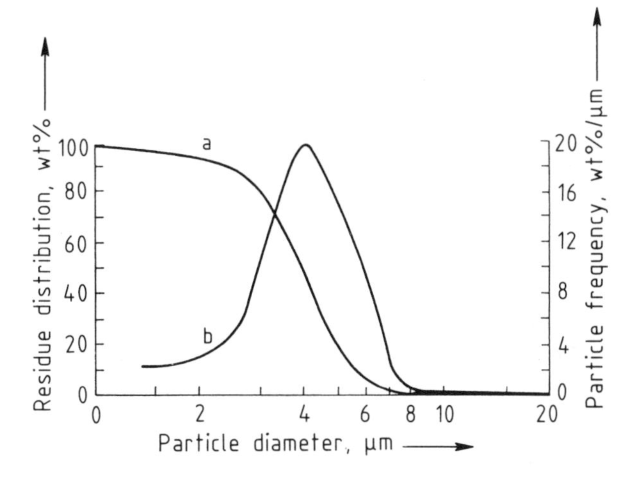

Figure 44. Residue summation curve and particle diameter distribution curve for zeolite 4A [87]
Coulter counter, 50 µm cell
a) Residue distribution;
b) Particle frequency

Additional differences result from the exceptional adsorption capability of the triphosphate anion [35].

The binary builder system sodium triphosphate–zeolite 4A assures outstanding multiple wash cycle performance; that is, incrustations resulting from precipitation of calcium and magnesium phosphates are largely avoided. Replacement of sodium triphosphate by zeolite 4A in a detergent leads to some deterioration in single wash cycle performance. By optimization of a detergent formulation with respect to surfactant, this effect can be compensated without difficulty.

Other sequestering agent carriers for zeolite 4A, such as NTA, and alternative ion exchangers like the water-soluble polycarboxylic acids have also achieved great importance for phosphate-free detergents.

Table 18. Comparison of various ion exchangers [7]

Substance (as sodium salt)	Calcium sequestration	Single wash cycle performance	Incrustation on fabrics and washing machines	Hygroscopicity	Biodegradability*
Poly(acrylic acid)	very high	good	little	very hygroscopic	minimal
Poly(α-hydroxyacrylic acid)	high	very good	little	hygroscopic	minimal
Poly(acrylic acid-*co*-allyl alcohol)	high	fairly good	little	very hygroscopic	minimal
Poly(4-methoxytetramethylene-1,2-dicarboxylic acid)	high	good	little	very hygroscopic	minimal
Poly(tetramethylene-1,2-dicarboxylic acid)	high	good	little	very hygroscopic	minimal
Poly(acrylic acid-*co*-maleic acid)	high	good	little	hygroscopic	minimal
Sodium aluminum silicate	high	good**	little	minimal	irrelevant

* Under the conditions of the Closed Bottle Test [246].

** In combination with appropriate soluble sequestering agents or soluble ion exchangers.

Zeolite 4 A, as it is mass produced for use in detergents, appears as cubic crystals with rounded corners and edges. These crystalline particles show a tendency to form agglomerates (Fig. 43).

The average diameter of a zeolite 4 A particle is ca. 4 μm when the maximum diameter is ca 20 μm. A diameter distribution curve obtained by differentiation of a residue summation curve (determined by using a Coulter counter) shows a sharp maximum at the corresponding abscissa value (Fig. 44).

Zeolite 4 A is specially optimized for use in detergents. Its unique particle form, cubes with rounded corners, protects fibers against damage, and the narrow range of particle sizes ensures that incrustation on laundry will be practically nonexistent.

Table 18 summarizes a number of important criteria for the evaluation of water-soluble and water-insoluble ion exchangers.

3.3 Bleaches

The term *bleach* can be taken in the widest sense to include the inducing of any change toward a lighter shade in the color of an object. Physically, this implies an increase in the reflectance of visible light at the expense of absorption.

In general, bleaching effects can occur through mechanical, physical, and/or chemical means, specifically through change or removal of dyes and soil adhering to the bleached object. In the washing process, all of these processes occur in parallel, but to varying extents. The relative significance of each is determined in part by the nature of the soil present. Mechanical/physical mechanisms are effective primarily for the removal of pigmented and greasy soil. Chemical bleaching is employed for the removal of nonwashable soils adhering to fibers and is accomplished by oxidative or reductive decomposition of chromophoric systems. Only oxidative bleaches are used to a great extent; many soils, commonly encountered in everyday laundry have been found to contain compounds which, if bleached reductively, become colorless but later return to their colored forms as a result of subsequent air oxidation. Nevertheless, this generalization does not rule out the use of special reductive bleaches (e.g., $NaHSO_3$, $Na_2S_2O_4$) to treat specific types of discoloration occurring in either household or institutional settings.

The extent of the bleaching effect that can be achieved is dependent on a number of factors, including the type of bleach, its concentration and residence time in the washing or rinsing process, the wash temperature, the type of soil to be bleached, and the nature of the fabric.

Bleachable soils encountered in household and institutional laundry consist of a broad spectrum of diverse substances [93], which are generally of vegetable origin and contain primarily polyphenolic compounds. These include the anthocyanin dyes (ranging from red to blue) derived from cherries, blueberries, and currants, and curcuma dyes from curry and mustard. The brown tannins found in fruit, tea, and wine stains arise from condensation of polyphenols with protein. Other brown organic polymers include the humic acids present, for example, in coffee, tea, and cocoa. The green dye chlorophyll and the red betanin from beets are pyrrole derivatives, as are the urobilin and urobilinogen derived from degradation of hemoglobin and discharged in urine. Carrot and tomato stains contain carotenoid dyes. Dyes of commercial origin such as those found in cosmetics, hair coloring agents, and ink are also important. Blood can also be regarded as a bleachable soil, but its removal can sometimes present problems [94].

Two procedures have attained major importance in oxidative bleaching during the washing and rinsing processes: peroxide bleaching and hypochlorite bleaching. The relative extent of their application varies, depending heavily on customs in particular countries.

3.3.1 Bleach-Active Compounds

The dominant bleaches in Europe are of the *peroxide* variety. Hydrogen peroxide is converted by alkaline medium to the active intermediate hydrogen peroxide anion according to the following equation:

$$H_2O_2 + OH^- \rightleftharpoons H_2O + HO_2^-$$

The usual sources of hydrogen peroxide are inorganic peroxides and peroxohydrates. The most important source is sodium perborate (sodium peroxyborate tetrahydrate, $NaBO_3 \cdot 4\,H_2O$) [95], which is present in crystalline form as the peroxodiborate ion:

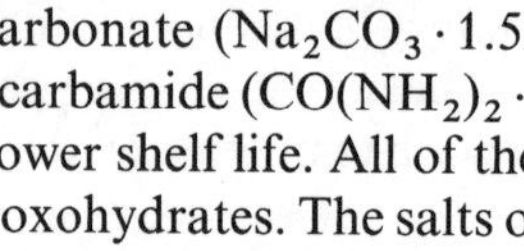

```
┌ HO     O–O     OH ┐2–
│   \   /   \   /   │
│     B       B     │
│   /   \   /   \   │
└ HO     O–O     OH ┘
```

This ion hydrolyzes in water to form hydrogen peroxide, but it also exhibits excellent shelf life when present in detergents. Attempts to replace sodium perborate in detergents by sodium percarbonate ($Na_2CO_3 \cdot 1.5H_2O_2$) [96], sodium perphosphate ($Na_4P_2O_7 \cdot 3H_2O_2$), percarbamide ($CO(NH_2)_2 \cdot H_2O_2$), etc. have proven to be problematic because of their lower shelf life. All of these compounds, in contrast to sodium perborate, are true peroxohydrates. The salts of peroxomono- and peroxodisulfuric acid and peroxomono- and peroxodiphosphoric acid are not significant as detergent bleaches. This results largely from their insufficient bleaching power in wash liquor, either because they fail to hydrolyze to hydrogen peroxide in an alkaline medium or because their oxidation potential is too low.

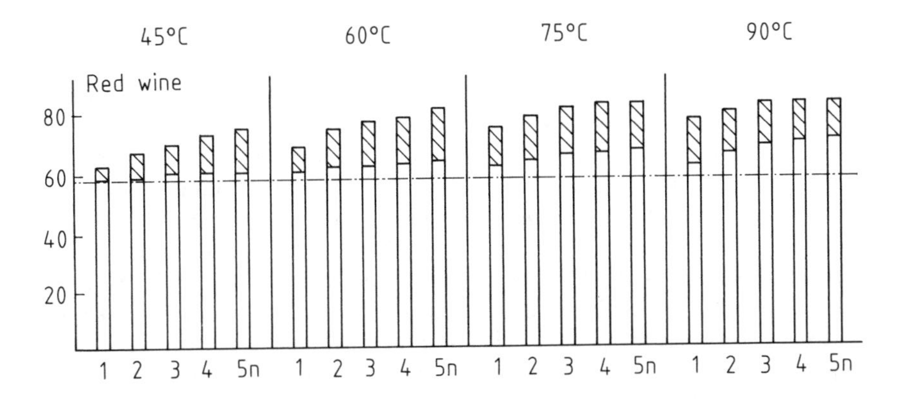

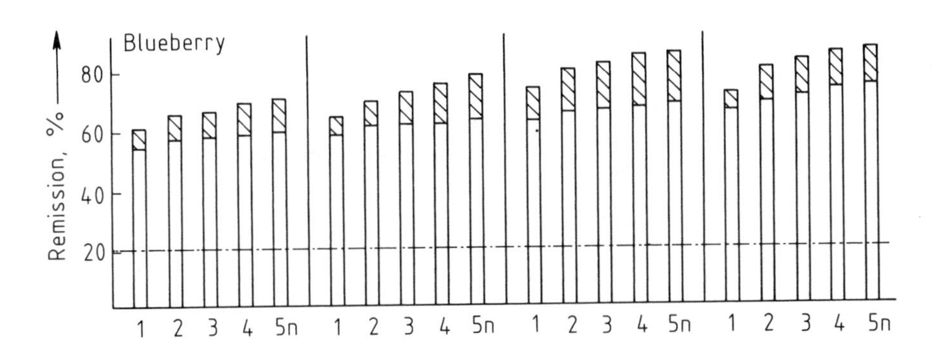

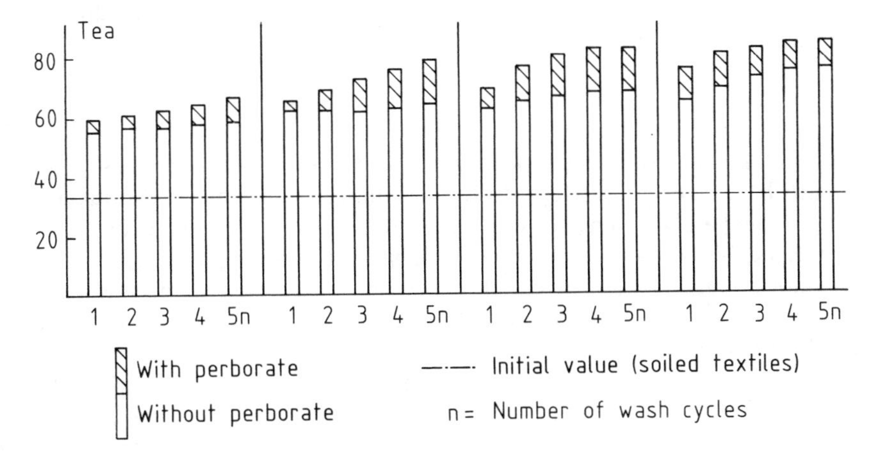

Figure 45. Bleach performance of sodium perborate vs. temperature [97]
Initial perborate concentration 1.5 g/L = 150 mg of active oxygen/L

Comparison shows that sodium perborate has the best prerequisites for use as a detergent bleach additive.

The concentration of bleach-active hydrogen peroxide anion increases with pH and temperature. Sodium perborate exhibits significantly less bleaching power below 60 °C, which requires application in washing machines with built-in heating coils. Such machines are common in Europe, but are rarely found in the United States or Japan. Even at low temperature (i.e., where the reaction equilibrium is unfavorable), hydrogen peroxide anions are present in the wash liquor, but under these conditions, they show only modest bleaching power (Fig. 45). The bleaching effect also increases markedly with increasing perborate concentration (Fig. 46).

Recently, various organic peroxy acids and their salts have begun to appear with sodium perborate in household detergents and laundry aids. With these it is possible to obtain significant bleaching at a temperature as low as 30 °C. Particularly worthy of mention are monoperoxyphthalic acid and diperoxydodecanedioic acid (DPDDA) salts:

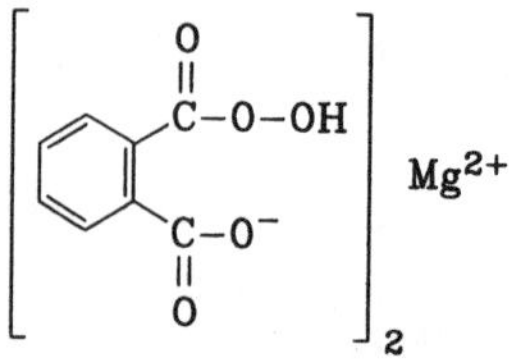

$$\left[C_6H_4(\text{C(=O)–O–OH})(\text{C(=O)–O}^-) \right]_2 Mg^{2+}$$

Monoperoxyphthalic acid monomagnesium salt

$$NaO{-}O{-}\overset{O}{\overset{\|}{C}}{-}(CH_2)_{10}{-}\overset{O}{\overset{\|}{C}}{-}O{-}ONa$$

Diperoxydodecanedioic acid sodium salt

Hypochlorite is used currently for bleaching in those countries where laundry habits cause sodium perborate to be less effective. In an alkaline medium, hypochlorite bleaches are converted to the hypochlorite anion:

$$HOCl + OH^- \rightleftharpoons ClO^- + H_2O$$

KAUFFMANN has concluded that hypochlorous acid is the active species [98], whereas AGSTER attributes bleaching effects to the decomposition of hypochlorous acid to hydrochloric acid and oxygen [99]. FLIS claims that both free hypochlorous acid and its anion, hypochlorite, are involved in the bleaching process [100].

Hypochlorite can be used in either the wash or the rinse at a concentration between ca. 50 and 400 mg/L active chlorine (Fig. 47) and a wash liquor ratio between 1:15 and 1:30.

Normally, an aqueous solution of sodium hypochlorite (NaOCl) is used as a hypochlorite source. Organic chlorine carriers (e.g., sodium dichloroisocyanurate), which hydrolyze to hypochlorite in an alkaline medium, are less common. Figure 48 shows a comparison of the bleaching properties of sodium hypochlorite and sodium perborate as a function of temperature and pH.

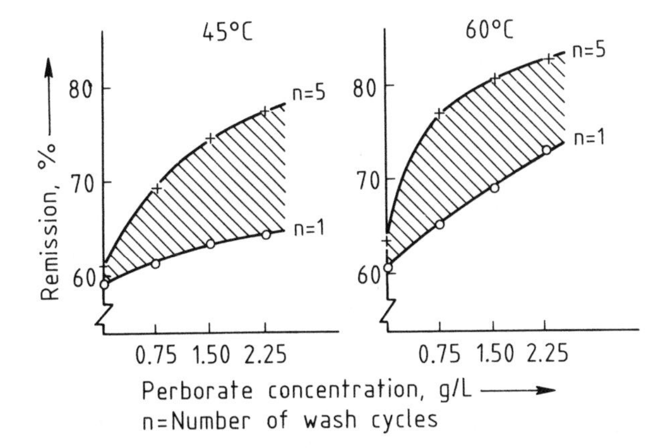

Figure 46. Bleach performance of sodium perborate vs. concentration (red wine soil) [97]

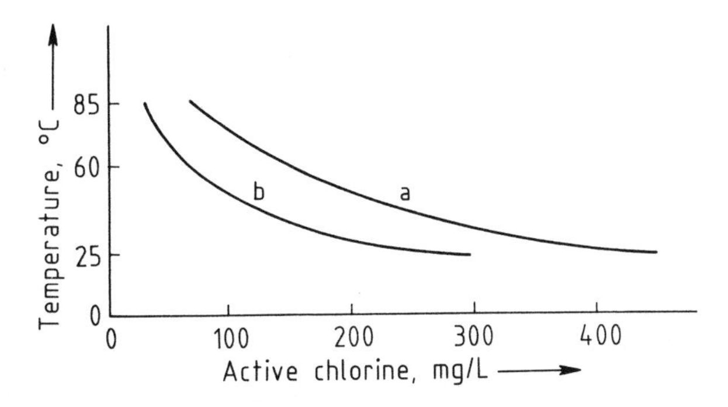

Figure 47. Recommended concentrations of active chlorine as a function of wash temperature [101]
a) Maximum concentration of application; b) Minimum concentration of application
"Active or available chlorine" is calculated as twice the actual weight percent of chlorine in the hypochlorite molecule

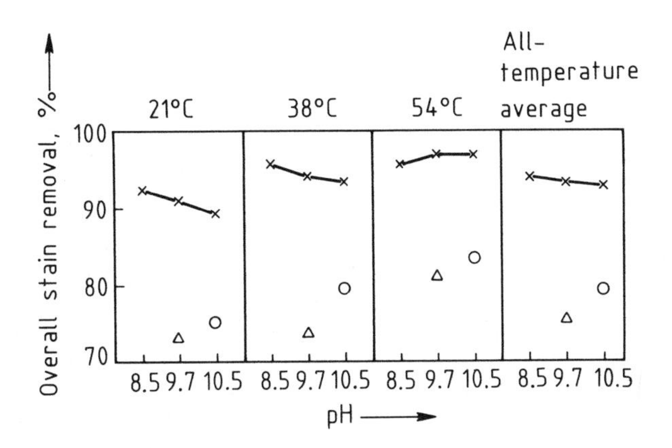

Figure 48. Stain removal comparisons [102]
× Hypochlorite (200 ppm active chlorine) + 0.15% detergent; ○ Perborate bleach (28 ppm active oxygen) with protease enzymes + 0.15% detergent; △ 0.15% Detergent only

Table 19. Washing conditions in different continents [104]

Washing conditions	United States/Canada	Japan	Western Europe
Washing machine	agitator type	impeller type	drum type
Heating coils	no	no	yes
Fabric load, kg	2–3	1–1.5	3–4
Amount of wash liquor, L	extra small: ca. 35 medium: ca. 50 large: ca. 65 extra large: ca. 80	low: 30 high: 45	low: 18–20 high: 25
Total water consumption, L (regular heavy cycle)	140	150	120
Wash liquor ratio	1:15–1:30	1:20–1:30	1:5–1:25
Washing time, min	10–15	5–15	60–70 (90 °C) 20–30 (30 °C)
Washing, rinsing, and spinning time, min	20–35	15–35	100–120 (90 °C) 40–50 (30 °C)
Washing temperature, °C	hot: 50 (122 °F) warm: 27–43 (80–110 °F) cold: 10–27 (50–80 °F)	10–40	90 60 40 30
Water hardness (median), ppm $CaCO_3$	relatively low, 100	very low, 50	relatively high, 250
Automatic detergent addition	mostly no	mostly no	dispenser
Recommended detergent dose*, g/L g/kg fabric	1.5 35–50	1.3 30–40	8–10 60–80
Peroxide bleach	+	(+)**	+
Chlorine bleach	+	(+)	(+)
Drying process by automatic dryer	+	(+)	(+)

* In the United States and Japan without bleaching components.
** Parentheses indicate that this element is less important.

One of the major advantages of sodium perborate over sodium hypochlorite is the fact that whereas the latter must be added separately in either the wash or the rinse cycle, perborate can be incorporated directly into a bleaching detergent. This results in the presence of a measured amount of a gentle peroxide bleach. By contrast, successful addition of chlorine bleach solution, whether in a household or an institutional machine, is heavily dependent on experience and adherence to the manufacturer's recommendations. Incorrect dosage of sodium hypochlorite can easily occur, and this can cause significant damage to laundry [103]. A further advantage of sodium perborate is its long shelf life, whereas sodium hypochlorite solutions have limited stability. On the other hand, hypochlorite bleaches can be used in both wash and rinse cycles without regard to temperature, and they provide effective bleaching at low temperature. However, because of its great reactivity and extraordinarily high oxidation potential, sodium hypochlorite, in contrast to sodium perborate, can lead to problems with textile dyes and fluorescent whitening agents, both of which often show poor stability to chlorine. Studies of washing and bleaching habits show that peroxide bleach use dominates in Europe, while hypochlorite bleach in either the wash or the rinse cycle is still the preferred bleaching agent in the rest of the world (cf. Table 19).

3.3.2 Bleach Activators

To achieve satisfactory bleaching with sodium perborate at temperatures < 60 °C, so-called bleach activators are commonly utilized. These are mainly acylating agents incorporated into detergents. When present in a wash liquor of pH 9–12, these activators react preferentially with hydrogen peroxide to form organic peroxy acids. As a result of their higher oxidation potentials relative to hydrogen peroxide, these intermediates demonstrate effective low-temperature bleaching properties. Such in situ peroxy acid bleaching agents are less aggressive with respect to fabric dyes and fluorescent whitening agents than sodium hypochlorite. Among the wide variety of bleach activators investigated [105], the following compounds have so far been incorporated into commercial products:

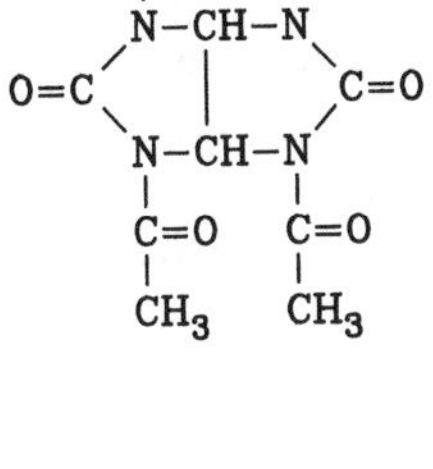

```
       CH3   CH3
       |     |
       C=O   C=O
       |     |
       N-CH-N
      /  |   \
  O=C    |    C=O
      \  |   /
       N-CH-N
       |     |
       C=O   C=O
       |     |
       CH3   CH3
```

Tetraacetylglycoluril (TAGU) [106],[107]

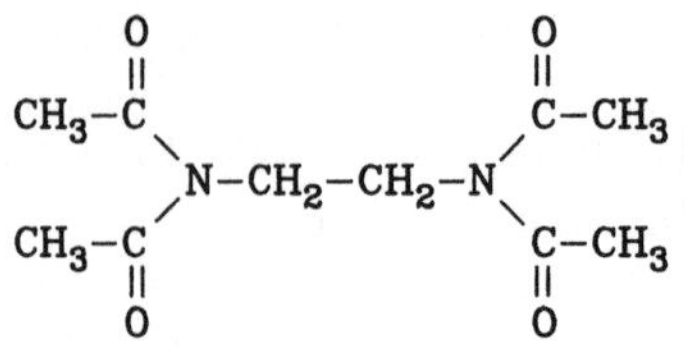

$$(CH_3-\overset{O}{\overset{\|}{C}})_2N-CH_2-CH_2-N(\overset{O}{\overset{\|}{C}}-CH_3)_2$$

Tetraacetylethylenediamine (TAED) [108],[109]

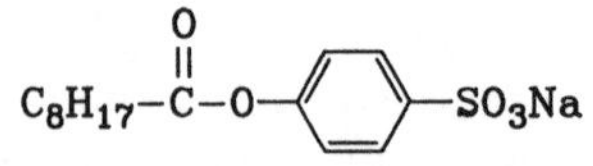

$$C_8H_{17}-\overset{O}{\overset{\|}{C}}-O-C_6H_4-SO_3Na$$

Sodium *p*-isononanoyloxy-benzenesulfonate (iso-NOBS) [110]

The first two are acylating agents that react with hydrogen peroxide according to the following scheme to produce peroxyacetic acid:

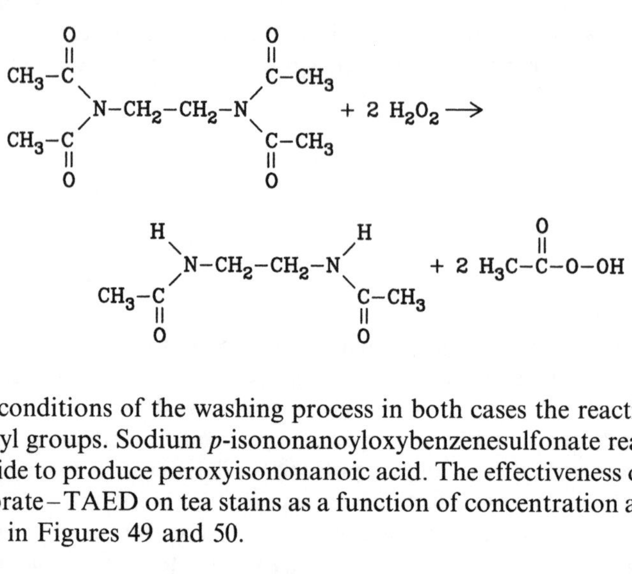

$$(CH_3-\overset{O}{\overset{\|}{C}})_2N-CH_2-CH_2-N(\overset{O}{\overset{\|}{C}}-CH_3)_2 + 2\ H_2O_2 \longrightarrow$$

$$CH_3-\overset{O}{\overset{\|}{C}}-NH-CH_2-CH_2-NH-\overset{O}{\overset{\|}{C}}-CH_3 + 2\ H_3C-\overset{O}{\overset{\|}{C}}-O-OH$$

Under the conditions of the washing process in both cases the reaction activates only two acetyl groups. Sodium *p*-isononanoyloxybenzenesulfonate reacts with hydrogen peroxide to produce peroxyisononanoic acid. The effectiveness of the system sodium perborate–TAED on tea stains as a function of concentration and temperature is shown in Figures 49 and 50.

3.3.3 Bleach Catalysts

Numerous attempts have been made to introduce into sodium perborate small amounts of catalysts that would increase its bleaching power, especially at low temperature. In most cases, traces of heavy metals have been employed. While addition of heavy metals alone can indeed cause decomposition of sodium perborate, it does not lead to a better bleach. In fact, bleaching action is usually diminished, and serious fiber damage occurs. The patent literature does describe a number of examples in which the bleaching power of sodium perborate can be

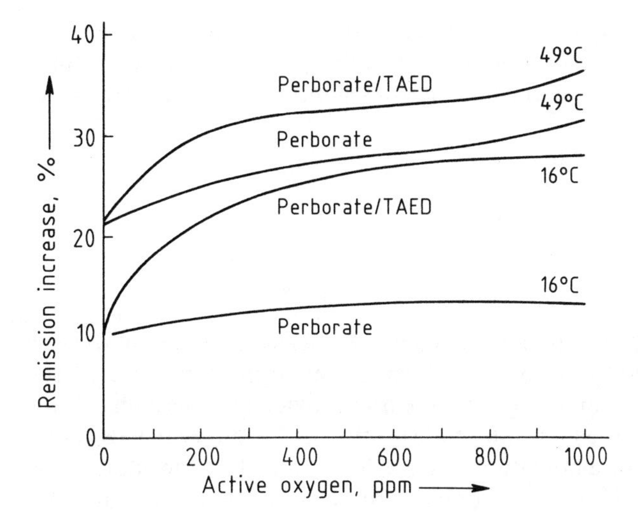

Figure 49. Bleaching effect as a function of bleach concentration [111]

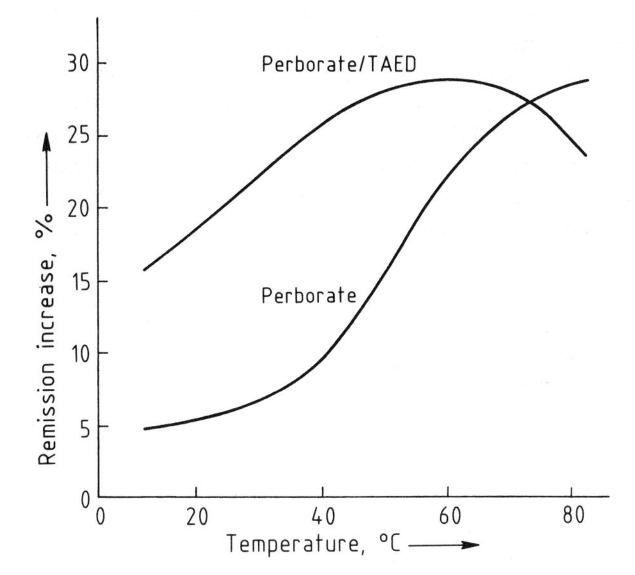

Figure 50. Bleaching effect as a function of temperature [111]

increased by heavy metal chelates [112]. However, the effectiveness of such systems remains controversial, and catalysts have not been accepted as a part of the washing process.

Photobleaching with aluminum or zinc phthalocyaninetetrasulfonate represents another form of bleach catalysis. With photobleaching, oxygen from the atmosphere is catalytically activated by a phthalocyanine compound, and the active oxygen in turn bleaches oxidizable stains. Suitable conditions for the few detergents that contain photobleaches include — apart from soaking in sunlight — slow drying of the laundry under conditions of high humidity and high light intensity. Therefore, photobleach catalysts are of interest only in countries subject to intense solar radiation.

3.3.4 Bleach Stabilizers

Traces of ions such as copper, manganese, and iron catalyze the release of oxygen from bleach systems. This reduces bleach effectiveness and at the same time causes damage to fabrics [113]. The addition of 0.1–5% finely divided magnesium silicate largely suppresses this catalysis as a result of absorption of the heavy metals [114].

An additional possibility for elimination of trace heavy metals is the addition of selective complexing agents. Through decomposition curves, Figure 51 shows the results of stability measurements on sodium perborate as a function of temperature and time. The detergent studied was based on 25% sodium perborate and 40% sodium triphosphate. Sodium triphosphate has only a modest complexing effect with heavy metals and is incapable of causing sufficient stabilization in their presence. By contrast, complexing agents like ethylenediaminetetraacetic acid (EDTA) or nitrilotrimethylenephosphonic acid (NTPO) (as well as other phosphonic acids) exhibit a marked stabilizing effect. Trace amounts of strong, selective complexing agents suffice. Especially favorable results are achieved with a combination of magnesium silicate and an effective complexing agent.

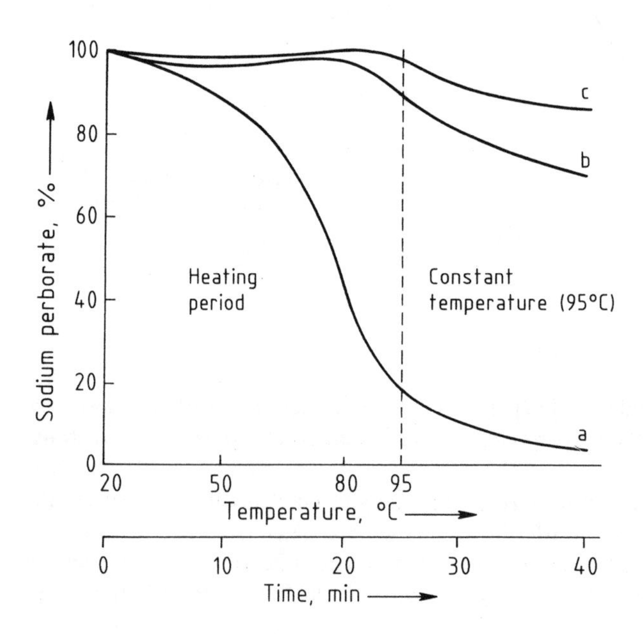

Figure 51. Stabilization of sodium perborate by complexing agents in the presence of traces of copper [7]
a) Detergent without heavy metal complexing agent;
b) Detergent with EDTA;
c) Detergent with NTPO

3.4 Auxiliary Agents

Surfactants, builders, and bleaches are quantitatively the major components of modern detergents; the *auxiliary agents* discussed in this section are introduced only in small amounts, each to accomplish its own specific purpose. Nevertheless, their absence from current detergent formulations is difficult to imagine.

3.4.1 Enzymes

Protein stains derived from sources such as milk, cocoa, blood, egg yolk, and grass are just as resistant to removal from fibers by simple detergents as are bleachable stains, particularly after the stains have dried. However, proteolytic (protein-cleaving) enzymes are usually capable of eliminating such soil without difficulty during the course of washing.

Use of enzymes in detergents was first described by OTTO RÖHM in a 1913 patent application [115]. Enzyme-containing detergents failed to play a major role in the following decades because the only available proteolytic enzyme preparation, a pancreatic extract obtained from slaughtered animals, was too sensitive to the alkaline and oxidative components of detergents. Only in the early 1960s was Röhm's idea subjected to intensive investigation, by which time it had become possible to prepare proteolytic enzymes by fermentation using specific strains of bacteria (*Bacillus subtilis*, later on *Bacillus licheniformis*). These enzymes were highly resistant to alkali and showed adequate stability at temperatures as high as ca. 65 °C for the time period required by normal wash processes. Commercial production of detergent enzymes experienced rapid expansion in the years that followed. For example, by 1969 nearly 80% of the detergents marketed in the Federal Republic of Germany contained enzyme additives. Industry subsequently reduced this proportion to < 50% as a result of public discussion of certain toxicological concerns. The reservations regarding detergent enzymes have since been addressed by technical modifications, particularly those involving the form in which enzymes are introduced into detergents: enzymes are now prepared as granulates, prills, or pellets [116]. As a result, more than 80% of the premium detergents produced in Western Europe again contain enzymes.

Progress has been less rapid in the United States, where only 20% of the detergents currently contain enzymes, but the trend is clearly toward their increased use.

A third major market for detergent enzymes is Japan. Only in the second half of the 1970s were proteases introduced into Japanese detergents, but since then enzyme-enhanced laundry products have shown rapid growth. Estimates show that more than 50% of the Japanese detergent market has been captured by enzyme formulations, and the use of enzymes in both powdered and liquid detergents continues to increase.

Table 20. Historical review of the preparation and use of detergent enzymes [117]

Year	Enzyme	Enzyme-containing detergents
1913	Otto Röhm claims the use of tryptic enzymes for detergents	detergents containing pancreatic enzymes (Röhm & Haas)
1927		optimized detergents containing pancreatic enzymes
1960	Alcalase	
post-1960	Microbial proteases made available on a commercial scale by Novo Industri, Copenhagen	first commercial product containing microbial proteases (presoak and wash pretreatment agent)
	Other producers followed, e.g., Maxatase from Gist en Spiritusfabrieken N.V., Delft; Nagase from Nagase Co.; Monlase 110 from Monsanto; Esperase from Novo	
1968		first heavy-duty detergent with microbial proteases
1969		microbial proteases contained in 80% of all detergents in the Federal Republic of Germany
1970		severe setback of addition of microbial proteases due to public criticism (the "allergy debate")
1972	Additional microbial enzymes suggested for use in detergents (amylases, lipases, pectinases, nucleases, oxidoreductases, etc.)	enzymes in detergents declared to be safe by the German Federal Health Agency
1975		market share of enzyme-containing detergents stabilizes in Germany at 80%

Table 20 provides a brief historical review of the production and application of detergent enzymes.

The effectiveness of enzymes is based on enzymatic hydrolysis of peptide and ester linkages. For an enzyme to be effective in the washing process, it must fulfill the conditions summarized in Table 21.

Serine-active, alkali-stable proteolytic enzymes constitute > 95 % of the enzymes used worldwide for detergent purposes. These bacillopeptidases (subtilisins) are distinguished by high temperature and pH stability (Fig. 52).

Only above a certain temperature does the stability of these enzymes decrease, and then very rapidly, leading to decomposition in a short time. Other modes of decomposition present little problem. For example, intensive research has shown

Table 21. Requisite properties for detergent enzymes [7]

Conditions during the wash process and storage	Enzyme requirements	Properties* of the principal types of microbial proteases			
		Serine proteases	Metallo-proteases	SH-proteases	Carboxy-proteases
Wash duration ≤1 h	high effectiveness				
Heterogeneous soil	low specificity				
Wash liquor temperatures to 95 °C	high temperature stability	+	++	○	+
Wash liquor pH between 9 and 11	stability to alkali	+	–	○	–
Wash liquor containing:					
sequestering agents	stability to sequestrants	+	–	+	+
perborate (H_2O_2)	stability to perborate	+	+	–	+
surfactants	stability to surfactants	○	○	○	○
Detergent storage over a period of months	little autolysis tendency				

* Key: ++ very high; + high; ○ fair; – low.

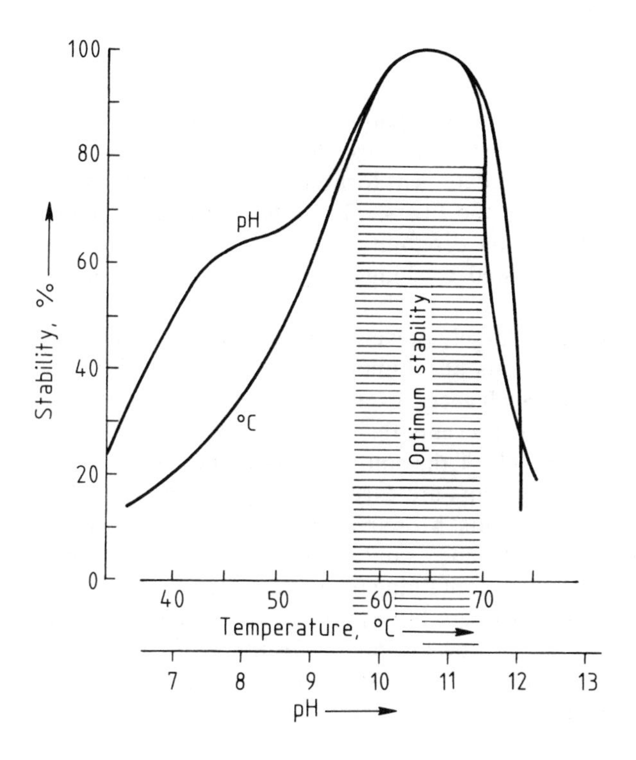

Figure 52. Conditions for optimum stability of Bacillopeptidase P 300 (Henkel) as a function of temperature and pH [117]

that, provided detergents containing them are properly formulated, the bacterial proteases currently available are not subject to significant damage by atmospheric oxygen either during storage or in the wash liquor. Stability problems with respect to anionic surfactants and sodium perborate were once the subject of debate, but usually these are readily solved by the proper choice of proteases and the appropriate formulation.

More recently, especially in the American and Japanese markets, amylases have been added to detergents along with proteases to take advantage of the activity of the amylases toward carbohydrate-containing soils.

In this context, the lipases which, due to their substrate specificity, should ease the removal of fat-containing soils are also worth mentioning. The activity of lipases is highly dependent on temperature and concentration. Studies of model greasy soils have shown that while removal of fats at a very low temperature (e.g., 20 °C) is largely due to the action of added lipases, at somewhat higher temperatures (e.g., 40 °C), surfactants, especially anionic surfactants, are largely responsible [118]. Overall, the advantages produced by lipase addition are minor under normal washing conditions. The stability of the usual commercial lipases under washing conditions is adequate, though not optimal.

3.4.2 Soil Antiredeposition Agents

The principal characteristic expected of a detergent is that it will cause soil to be removed from textile fibers during the washing process. Removed soil is normally finely dispersed, and if a less than optimal detergent formulation is employed, some or all of it may at some point return to the fibers. This is termed a wash liquor showing "insufficient soil antiredeposition capability" [119], [120]. The problem becomes especially apparent after repeated washing as a distinct graying of the laundry.

Redeposition of displaced soil can be largely prevented by carefully choosing the various detergent components (surfactants and builders), but addition of special antiredeposition agents is also helpful. Such agents exercise their effects by becoming adsorbed irreversibly, i.e., in a way that prevents their removal by water, on both textile fibers and soil particles. Approach of the soil to fibers is thereby hindered [121], [122]. Classical antiredeposition agents are carboxymethyl cellulose (CMC) derivatives bearing relatively few substituents. More recently, analogous derivatives of carboxymethyl starch (CMS) have played a similar role. Unfortunately, these substances are effective only with cellulose-containing fibers such as cotton. With the increasing replacement of these natural fibers by synthetics, on which CMC has virtually no effect, the need has arisen to develop other effective antiredeposition agents. Some surfactants have been found to be well-suited to the purpose (Fig. 53), as are such nonionic cellulose ethers [123], [124] as the following:

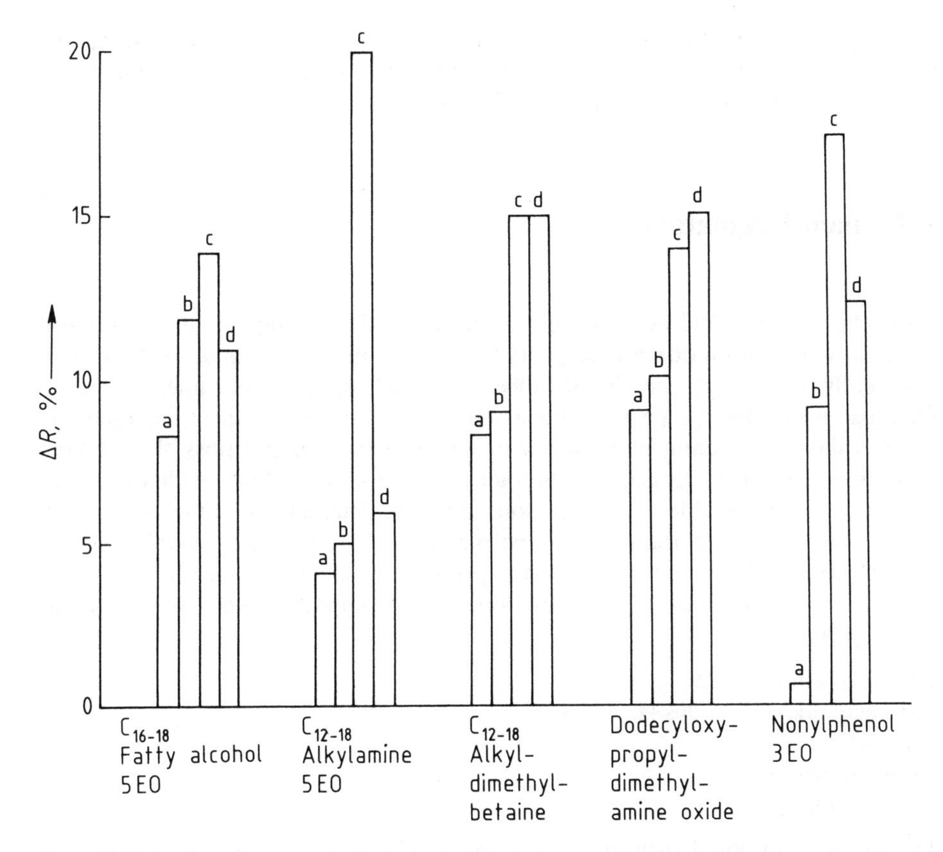

Figure 53. Influence of specific surfactants on the soil antiredeposition capability of detergents [43]
ΔR indicates the improvement in the soil antiredeposition capability conferred by a given additive
a) Cotton, 95 °C; b) Resin-finished cotton, 95 °C; c) Polyester, 60 °C; d) Polyester/cotton, 60 °C
5 g/L Detergent; 0.2 g/L surfactant; number of wash cycles: 3; time: 30 min

R:
- $-CH_3$
- $-CH_2CH_3$
- $-CH_2CH_2OH$
- $-CH_2CHOHCH_3$
- $-CH_2CH_2CHOHCH_3$

Because of the multitude of fabrics currently on the market, manufacturers supplement most modern detergents with mixtures of anionic and nonionic polymers (e.g., carboxymethyl cellulose–methyl cellulose).

3.4.3 Foam Regulators

In the days of soap detergents, foam was understood as an important measure for washing power. With modern detergents based on synthetic surfactants, foam has lost virtually all of its former significance. Nonetheless, most consumers — apart from those using drum-type washing machines — still expect their detergent to produce voluminous foam, preferably comprised of the smallest possible bubbles. The reasons seem to be largely psychological (e.g., foam provides evidence of detergent activity and it hides the soil). Consequently, detergents designed for use in washing machines other than drum-type machines (i.e., products mainly for the non-European market) often are caused to produce the desired foam characteristics by addition of small amounts of foam boosters. Compound types suited to the purpose include

fatty acid amides [125]
fatty acid alkanolamides [126], [127]
betaines
sulfobetaines [128]
amine oxides [129]

In Europe, household drum-type washing machines are becoming increasingly common; with such machines, only weakly or moderately foaming detergents are permissible. Thus, foam boosters have lost their former significance in the European market, with the exception of the role they retain in certain specialty detergents (e.g., detergents for hand washables). Especially at high temperature, heavy foaming can cause overflowing in drum-type machines, often accompanied by considerable loss of active ingredients. Furthermore, large amounts of foam reduce the mechanical action to which laundry is subjected in such machines. For these reasons, foam regulators — often somewhat incorrectly described as "foam inhibitors" — are commonly added to minimize detergent foaming tendencies [130]–[133].

Ensuring effective foam regulation requires that the regulating system be precisely matched to the other detergent components present. The most difficult cases are surfactant combinations with exceptionally high foam stability; by contrast, an unstable foam, even if present in large amounts, seldom causes a problem. The principal requirements for a substance to show foam regulating capability are extremely low water solubility and high surface spreading pressure. Foam regulators show a wide range of physicochemical properties, but their mechanism of action can usually be explained by assuming that they either force surfactant molecules away from boundary surfaces or they penetrate boundary surfaces that are already occupied by surfactants, thereby creating defects. These defects weaken the mechanical strength of the foam lamellae and cause their rupture.

A great many foam regulators are described in the patent literature, but relatively few have had any real impact. The detergents on the market that are based on LAS/fatty alcohol polyglycol ethers are effectively controlled by soaps with a broad chain length spectrum (C_{12-22}) [134], [135]. Such soaps have limitations, however, for the following reasons:

1) The foam-depressing activity of soaps is largely due to calcium salts that arise during the washing process, with the calcium originating either as hard water or as calcium-containing soil. In other words, satisfactory foam regulation occurs only if a sufficiently high calcium ion content is assured. This is absent if exceptionally soft water is used and if the laundry is only lightly soiled, and foaming problems can result.

2) The foam regulating power of soap is substantially lower with detergents based on anionic surfactants other than alkylbenzenesulfonates (e.g., α-olefinsulfonates, fatty alcohol sulfates, α-sulfo fatty acid esters).

3) If soap is used as a foam regulator, the only complexing agents that can be added as builders are those with limited stability constants; i.e., even in the presence of builders, the calcium ion concentration must remain high enough to permit formation of sufficient lime soap to act as a foam regulator. An advantage of sodium triphosphate, widely used as a complexing agent in detergents, is that stability constant values for calcium triphosphate complexes are quite favorable. Many other complexing agents (e.g., nitrilotriacetic acid, NTA) have higher stability constants, but these prevent the formation of foam-regulating lime soaps (Fig. 54).

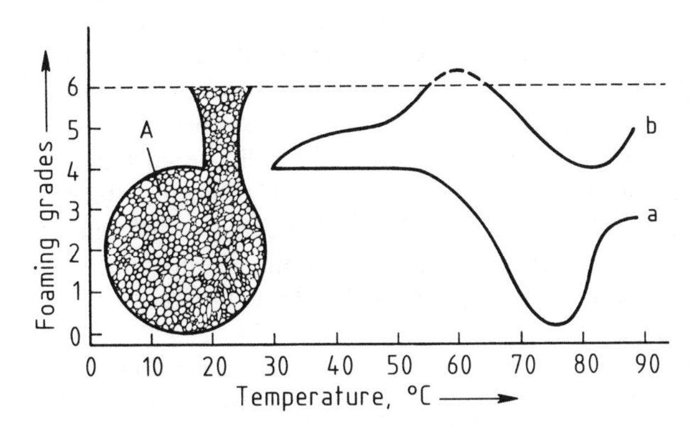

Figure 54. Foaming behavior in a drum-type washing machine [7]
A) Schematic drawing of the inner drum of a drum-type washing machine, including dispenser
a) 7.5 g/L Detergent based on 40% sodium triphosphate;
b) 7.5 g/L Detergent based on 40% NTA
Foaming grades: 0, no foam; 4, foam at the upper edge of the sight glass; 5, foam in the dispenser; 6, foam overflow

Recent investigations have shown that specific silica-activated silicone or paraffin oil systems are considerably more universal in their applicability and that their effectiveness is independent of both water hardness and the nature of the surfactant–builder system employed [136]–[138].

3.4.4 Corrosion Inhibitors

Washing machines currently on the market are constructed almost exclusively with drums and laundry tubs of corrosion-resistant stainless steel or with an enameled finish that is inert to alkaline wash liquors. Nevertheless, various machine components are made of metals such as aluminum, especially in older models. To prevent corrosion of these parts, modern detergents are supplemented with corrosion inhibitors in the form of water glass. The colloidal silicate that is present deposits as a thin, inert layer on metallic surfaces, thereby protecting them from attack by aqueous hydroxide ions.

3.4.5 Fluorescent Whitening Agents

Properly washed and bleached white laundry, even when clean, actually has a slight yellow tinge. For this reason, as early as the middle of the 19th century, people began treating laundry with a trace of blue dye (blueing agents, e.g., ultramarine blue) so that the color was modified slightly and a more intense visual sensation was produced. Modern detergents contain fluorescent whitening agents (FWA), called *optical brighteners,* to accomplish the same purpose [93]. Fluorescent whitening agents are organic compounds that convert a portion of the invisible ultraviolet radiation in sunlight into longer wavelength blue light. It is well-known that the yellowish cast of freshly washed and bleached laundry is a result of partial absorption of the blue radiation reaching it, resulting in reflected light that is deficient in the blue region of its spectrum. The radiation emitted by whitening agents makes up for this deficiency, so that the laundry becomes both brighter and more white (Fig. 55).

This process of compensating for differences in yellow and blue radiation is fundamentally different from treatment with a blueing agent (washing blue). The latter entails subtractive absorption of yellow light, which results in an overall reduction in brightness (Fig. 56).

Limits exist for the extent of whitening that can be achieved by fluorescent whitening agents. This is due in part to the fact that these agents are themselves dyes that exhibit a certain amount of reflectance in the visible region in addition to their emission. Thus, any agglomeration or buildup of fluorescent whitening agent on the

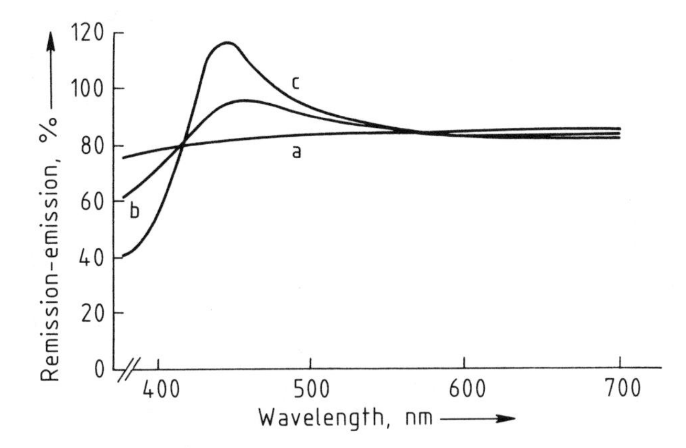

Figure 55. Remission–emission curve of a bleached fabric after application of a fluorescent whitening agent [39]
a) Bleached fabric; b) Fabric treated with small amounts of whitener; c) Fabric treated with large amounts of whitener

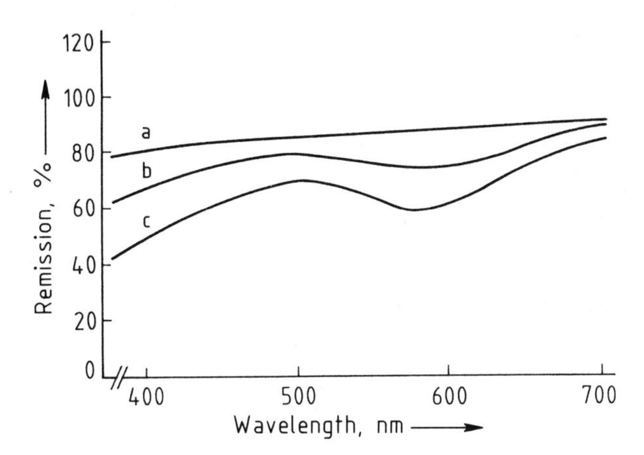

Figure 56. Remission curve of a bleached fabric after application of ultramarine blue [39]
a) Bleached fabric; b) Weakly blued fabric; c) Intensely blued fabric

fibers eventually becomes apparent. Furthermore, the absorption edge gradually shifts into the visible region of the spectrum, leading to emission at a higher wavelength. These factors can work in combination with the innate color of the fabric and eventually result in a visible change of shade (discoloration).

Fluorescent whitening agents can be divided into the following categories, depending on their applicability to detergent use:

cotton whiteners
chlorine-resistant whiteners
polyamide whiteners
polyester whiteners

The major commercial products are based on five basic structural frameworks [139]: stilbene, diphenylstilbene, coumarin–quinolone, diphenylpyrazoline, and the combination of benzoxazole or benzimidazole with conjugated systems. Table 22

Table 22. Summary of key fluorescent whitening agents [93]

General basic structures	Substituents	Substrate *
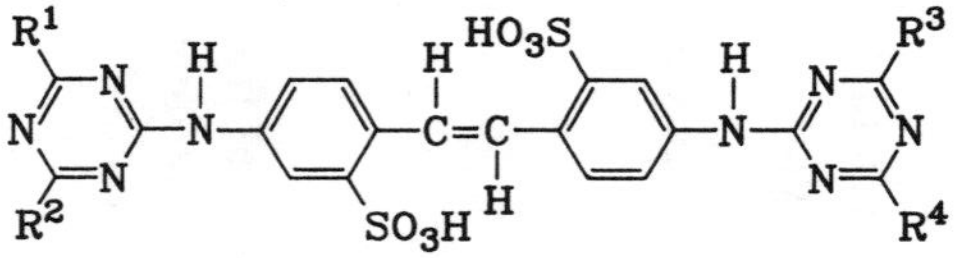 4,4'-Bis(triazinylamino)stilbene-2,2'-disulfonic acids	R 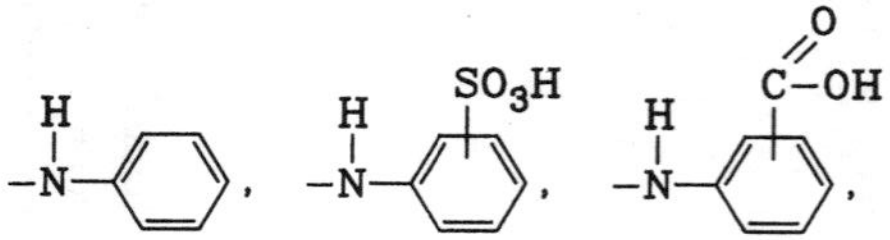 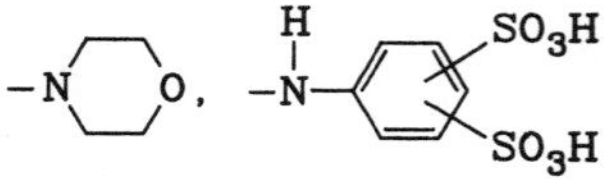 = anilino ($-NH-C_6H_5$), sulfoanilino ($-NH-C_6H_4-SO_3H$), carboxyanilino ($-NH-C_6H_4-COOH$), morpholino, disulfoanilino ($-NH-C_6H_3(SO_3H)_2$), $-NH-(CH_2)_{2-3}-OCH_3$, $-N(CH_2-CH_2-OH)_2$, $-N(CH_3)-CH_2-CH_2-OH$, $-NH_2$, $-NH$-alkyl, $-N(alkyl)_2$, $-OCH_3$, Cl, $-NH-CH_2-CH_2-SO_3H$, $-NH-CH_2-CH_2-OH$	CO, LI VI, MD PA WO
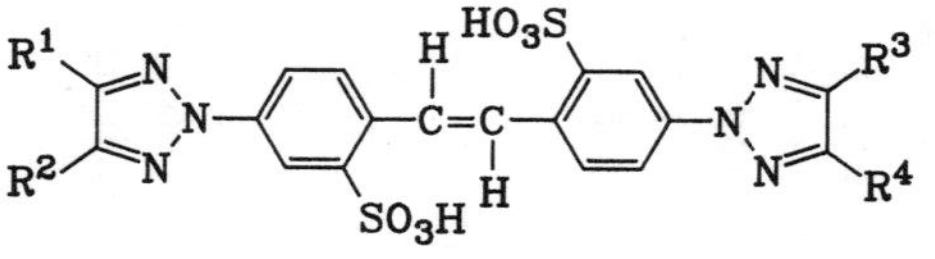 4,4'-Bis(*v*-triazol-2-yl)stilbene-2,2'-disulfonic acids	R = H, alkyl, 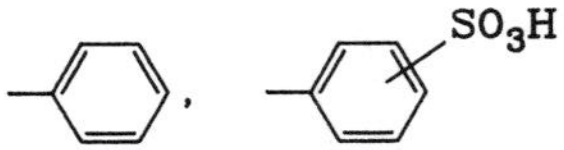 phenyl ($-C_6H_5$), sulfophenyl ($-C_6H_4-SO_3H$)	CO, LI VI, MD PA
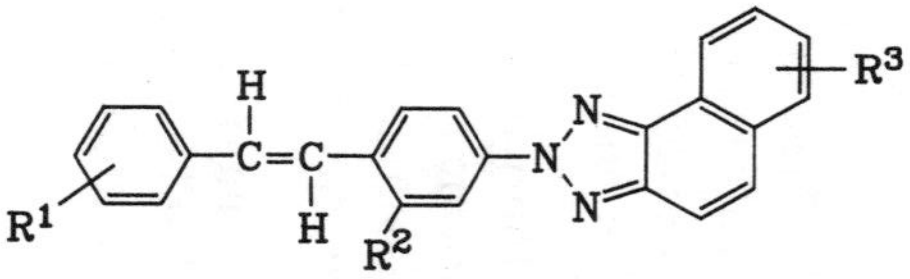 Stilbenylnaphthotriazoles	R^1 = H, Cl, $-NH-CH_3$, $-N(CH_3)_2$ R^2 = $-SO_3H$, $-SO_2-N\langle$, $-SO_2-OC_6H_5$, $-CN$ R^3 = H, $-SO_3H$, $-OCH_3$	CO, LI VI, MD PA WO, SI

Table 22. (continued)

General basic structures	Substituents	Substrate*
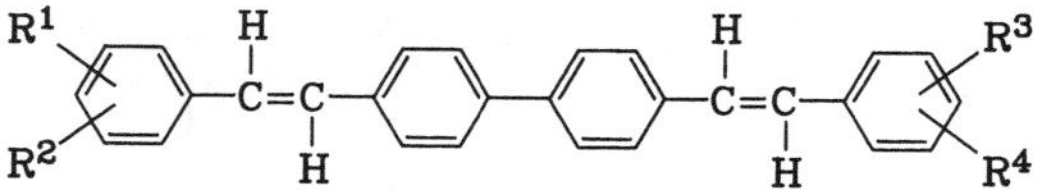 4,4'-Bis(styryl)biphenyls	R = H, $-SO_3H$, $-SO_2-N(alkyl)_2$, $-OCH_3$, $-CN$, Cl, $-C(=O)-OCH_3$, $-C(=O)-N(alkyl)_2$	CO, LI VI, MD PA
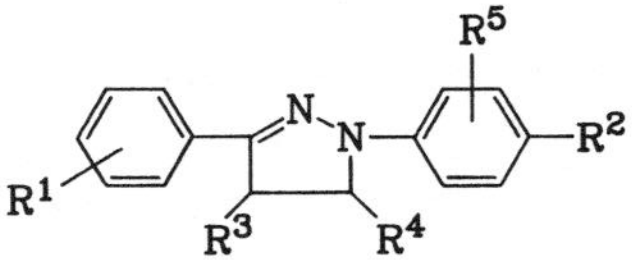 Pyrazolines	R^1 = H, Cl, $-N<$ R^3, R^4 = H, alkyl, 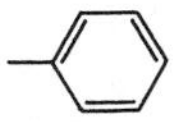 phenyl R^2 = Cl, $-SO_3H$, $-SO_2-NH_2$, H, $-SO_2-NH-$, $-C(=O)O-alkyl$, $-C(=O)O-(CH_2)_{2-3}-O-alkyl$, $-SO_2-CH_3$, $-SO_2-CH_2-CH_2-OH$, $-SO_2-CH=CH_2$, $-SO_2-NH-CH_2-CH_2-CH_2-N^+(CH_3)_3$ R^5 = H, Cl	AC, TA WO, SI PC PA
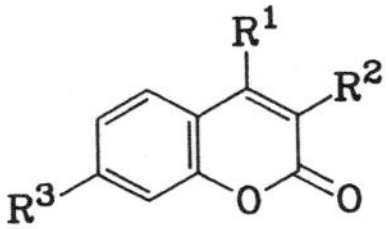 Coumarins	R^1 = H, $-CH_3$, $-CH_2C(=O)-OH$ R^2 = H, 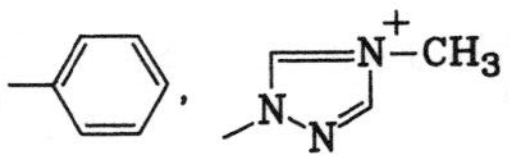 phenyl, N-methyltriazolium ($-N^+-CH_3$), $-C(=O)-OCH_3$ R^3 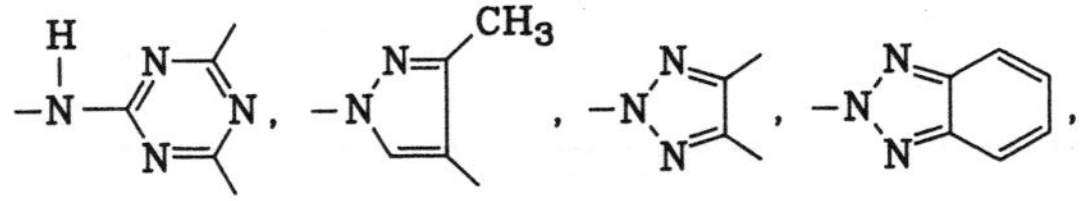 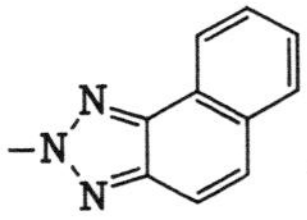 = $-NH-$(dimethyltriazinyl), $-N$(3,4-dimethylpyrazolyl) (CH_3), $-N$(dimethyltriazolyl), $-N$(benzotriazolyl), $-N$(naphthotriazolyl), $-O-alkyl$, $-N(alkyl)_2$, $-N(H)-C(=O)-CH_3$	AC, TA PA PE WO, SI

Table 22. (continued)

General basic structures	Substituents	Substrate*
Quinolones	R^1 = H, $-CH_3$, $-C_2H_5$ R^2 = H, phenyl R^3 = alkyl, $-NH-CH_3$, $-N(CH_3)_2$	AC, TA PA WO, SI
Bis(benzoxazol-2-yl) derivatives	X = thiophene-2,5-diyl, furan-2,5-diyl, naphthalene-1,4-diyl, stilbene-4,4′-diyl, $-CH=CH-$, $-CH=CH-C_6H_4-$ R = H, $-C(CH_3)_3$, $-C(CH_3)_2-C_6H_5$, $-CH_3$, $-CO-O-$alkyl	AC, TA PA PE CL PP
Bis(benzimidazol-2-yl) derivatives	X = furan-2,5-diyl, thiophene-2,5-diyl, $-CH=CH-$ R = H, $-CH_3$, $-CH_2-CH_2-OH$	AC, TA PA CL PP

Table 22. (continued)

General basic structures	Substituents	Substrate*
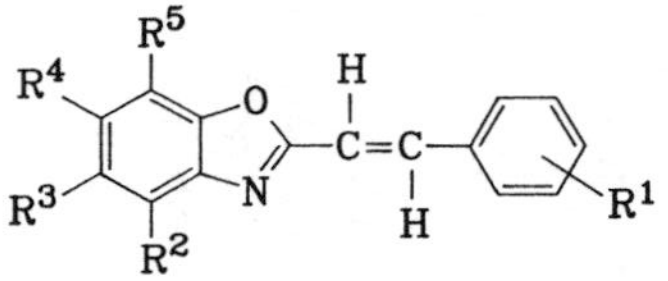 2-Styrylbenzoxazoles and -naphthoxazoles	R^1 = –CN, –C(=O)O-alkyl, Cl R^2, R^3 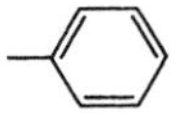 = –phenyl R^3, R^5 = H, alkyl R^4 = H, alkyl, 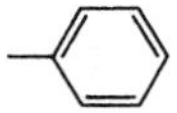 –phenyl	AC, TA PA PE

*CO Cotton, LI Linen, VI Viscose, MD Modal, AC Acetate, TA Triacetate, PA Polyamide, PE Polyester, CL Poly(vinyl chloride), PC Polyacrylonitrile, PP Polypropylene, WO Wool, SI Silk

provides an overview of the most important cases, including information regarding their affinities to specific types of fibers during the washing process.

The application of FWA in the wash liquor is essentially a dyeing process. In the case of cotton and chlorine-resistant FWA, binding occurs through the formation of hydrogen bonds to the fibers. Resin-finished cotton is generally less susceptible to the effects of FWA. Whitening effects achieved with polyamide whiteners, and particularly the polyester whiteners (which tend to be poorly adsorbed), are due largely to the diffusing power of whitening agent molecules present at the fiber surfaces.

Fluorescent whitening agents are evaluated not only on the basis of their affinity for the relevant fibers, but also on their stability and fastness. "Stability" means their resistance to chemical change in the course of the washing process and prior to their adsorption on fibers. "Fastness" refers to their chemical resistance after adsorption. Of primary concern is fastness against light and oxygen, as well as good oxygen stability and, in countries where chlorine bleach is used, fastness and stability with respect to chlorine.

Virtually all detergents currently on the market contain fluorescent whitening agents. Nevertheless, their use on certain pastel fabrics can cause unwanted color changes; for this reason special detergents that lack FWA are also offered.

3.4.6 Fragrances

Fragrances were first added to detergents in the 1950s. Their presence is more than simply a fad or a matter of fashion. Thus, apart from their role in providing detergents with an agreeable odor, an important function of fragrances is to mask unpleasant odors arising from the wash liquor during washing. This becomes particularly important as more washing machines find their way into the living areas of homes. Fragrances are also intended to confer a fresh, pleasant odor on the laundry itself [140]–[143].

Detergent fragrances are generally present only in very low concentrations (< 1%), but they tend to be extremely complex mixtures (Table 23).

Several factors are involved in establishing the composition of a fragrance mixture apart from odor and the cost of the often expensive individual components. In particular, the detergent formulation must be taken into careful account, as must the properties of the fabrics to be washed. Chemical stability relative to other detergent ingredients is especially important, as is the limited volatility of the individual fragrances. Temperatures to which the detergent will be exposed during storage must also be considered.

The fragrance industry currently has at its disposal for the preparation of detergent fragrance oils not only synthetic substances, but also a wide range of natural fragrances. The overall palette is, nevertheless, restricted considerably due to the instability of many fragrances and the fact that some have a tendency to discolor detergents to which they are added.

Table 23. Hypothetical formulation of a fragrance mixture for detergent use [143]

Perfuming agent	Quantity, g
Isoeugenol	10
Undecylenaldehyde	10
Benzyl acetate	20
Allylionone	20
Vigorose (Naarden)	20
Dimetol (Givaudan)	20
Ionone, pure	20
Coumarin	20
Ocimenyl acetate	20
Anisaldehyde	30
Cinnamyl alcohol	30
Terpineol	50
Dihydromyrcenol	50
Floramat (Henkel)	50
Vertacetal (Dragoco)	50
p-tert-Butylcyclohexyl acetate	50
Cyclamenaldehyde	50
Mugaflor (Haarmann & Reimer)	50
Citronellol	50
Galaxolide 50 (IFF)*	50
Benzyl salicylate	50
Linalool	80
Phenethyl alcohol	100
α-Amylcinnamaldehyde	100
Total	1000

* International Flavors and Fragrances.

3.4.7 Dyes

Until the 1950s, powdered detergents were more or less white, consistent with the color of their components. Thereafter, products were commonly encountered in which colored granules were present along with the basically white powder: certain components had been deliberately dyed to make the products more distinctive. In the meantime, uniformly colored detergents have also appeared on the market, and the idea of introducing coloring agents has become quite common. The preferred colors for both powders and liquids are blue, green, and pink.

There are two important criteria for selecting a coloring agent:

1) Good storage stability with respect to other detergent components and to light
2) No significant tendency to affect textile fibers

3.4.8 Fillers and Formulation Aids

The usual *fillers* for powdered detergents are inorganic salts, especially sodium sulfate. Their purpose is to confer the following properties on a detergent:

flowability
good flushing properties
high solubility
no caking of the powder even under highly humid conditions
no dusting

Formulation aids are substances required in the preparation of liquid detergents. The most important of these have the assignment of ensuring through their own hydrotropic characteristics that the other detergent components can be combined in a stable way in an aqueous environment. Above all, these ingredients must prevent phase separation und precipitation occurring as a result of shifts in temperature. Commonly used materials include short-chain alkylbenzenesulfonates (e.g., toluenesulfonate and cumenesulfonate) and urea, as well as low-molecular mass alcohols (ethanol, 2-propanol) and polyglycol ethers (poly(ethylene glycol)).

4 Household Detergents

Detergents currently on the market in various parts of the world can be classified into the following groups from a detergency standpoint:

heavy-duty or all-purpose detergents
specialty detergents
laundry aids
aftertreatment aids

4.1 Heavy-Duty Detergents

The category *heavy-duty detergents* includes those detergent products suited to all types of washing und usually to all wash temperatures. They are offered in both powdered and liquid form. Nevertheless, great differences are found from one formulation to another. For example, ecological factors induce manufacturers to offer in different countries or regions powdered products both with and without sodium triphosphate. Figure 57 provides an overview of the market share of individual heavy-duty detergent types.

4.1.1 Powdered Heavy-Duty Detergents

Significant composition differences exist among available powdered heavy-duty detergents around the world. An approximate breakdown of detergent formulations for various countries and continents is provided in Table 24.

A more detailed indication of formulation ranges for high-quality heavy-duty detergents in the European, United States, and Japanese markets is presented in Table 25.

Detergent doses in Europe generally fall between 6 and 10 g of detergent per liter of wash liquor, whereas in the United States, Japan, and Brazil, values of 1 – 1.5 g/L are more typical. There are a number of reasons for the latter being so much lower, including differences in washing machines, softer water, 3-fold higher wash liquor ratios, separate addition of bleach, e.g., hypochlorite solution instead of sodium

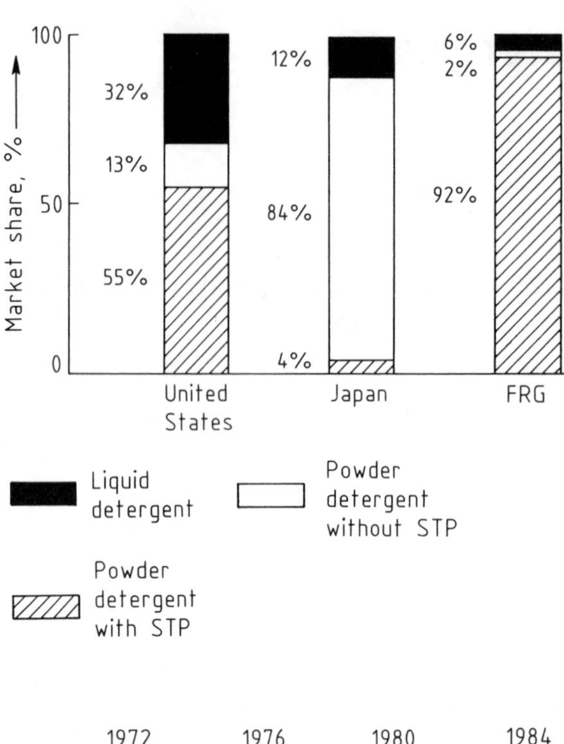

Figure 57. Market share of different heavy-duty detergent types (1985) [104]
STP Sodium triphosphate

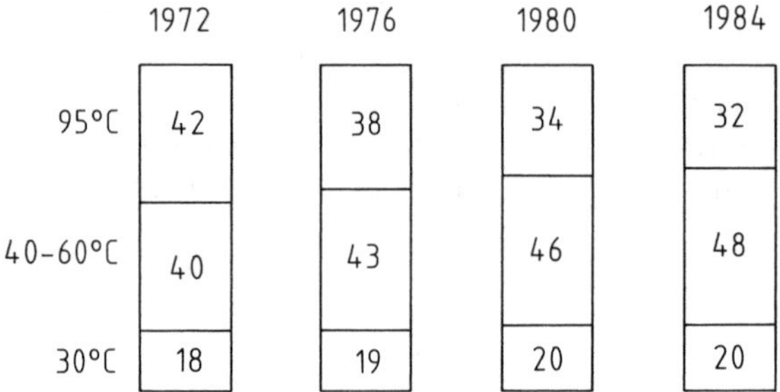

Figure 58. Percentage of wash loads vs. washing temperature in the Federal Republic of Germany (in %) [104]
Numbers below 30 °C have not been considered. Source: Henkel market research

perborate, and different detergent formulations [145]. Detergents in the United States, Japan, and Brazil are designed for agitator-type washing machines (cf. Chap. 13), and these detergents lack foam inhibitors; however, they often contain foam boosters instead.

The sudden rise in popularity of brightly colored and easy-care fabrics in the last 10–15 years, along with the trend toward energy conservation, has resulted in a general decline in the once common European practice of washing at 95 °C, while a 40–60 °C wash has gained favor. The change is illustrated in Figure 58 for the time period 1972–1984 in the Federal Republic of Germany. A similar trend from hot ($\geq$ 50 °C) washing to warm (27–43 °C) is apparent in the United States (Fig. 59).

Table 24. General formulations for powdered heavy-duty detergents [144]

Ingredients	Composition, %			
	United States, Canada, Australia	South America, Middle East, Africa	Europe	Japan
Actives[a]	8–20	17–32	8–13	19–25
Foam boosters[b]	0–2		0–3	
Foam depressants[c]			0.3–5	1–4
Builders[d]				
Sodium triphosphate	25–35	20–30	20–35	0–15
Mixed or nonphosphate	15–30	25–30	20–45	0–20
Sodium carbonate	0–50	0–60		5–20
Antiredeposition agents[e]	0.1–0.9	0.2–1	0.4–1.5	1–2
Anticorrosion agents[f]	5–10	5–12	5–9	5–15
Optical brighteners	0.1–0.75	0.08–0.5	0.1–0.75	0.1–0.8
Bleach[g]	[h]		15–30	0–5
Enzymes[i]	[h]		0–0.75	0–0.5
Water	6–20	6–13	4–20	5–10
Fillers[j]	20–45	10–35	5–45	30–45

[a] Mostly alkylsulfonates (linear and branched), fatty alcohol ethoxylates, and fatty alcohol sulfates.
[b] Ethanolamides, such as coco monoethanolamide.
[c] In small levels, silicones; at higher levels, soaps.
[d] Some formulations use only sodium triphosphate; others have mixtures of triphosphate with other phosphates (for example, sodium orthophosphate), sodium aluminum silicates (zeolites), the sodium salt of nitrilotriacetic acid, sodium citrate, and sodium carbonate; still others use the other builders alone or in various combinations.
[e] Sodium carboxymethyl cellulose, other cellulose-based polymers, or synthetic polymers.
[f] Sodium silicates (as purchase solutions).
[g] In the United States, sodium perborate, when used. In Europe, formulations for high-temperature washing have sodium perborate, whereas those for use at lower temperatures include boosters such as tetraacetylethylenediamine.
[h] A few formulations include 20–25% bleach and up to 0.75% enzymes.
[i] As purchased granules.
[j] Predominantly sodium sulfate.
[k] Usually also include small amounts of dyes, opacifiers if desired, fragrances, etc.

Detergent manufacturers have responded to these changes by increasing use of nonionic surfactants, noted for their effectiveness at low temperature, as well as by adding enzymes and bleach activators.

Dual-function detergents represent a unique development originating in the United States. These products, in addition to fulfilling their primary function of providing effective cleansing, also contain materials designed to impart specific desirable characteristics on the laundry; e.g., they may act as fabric softeners. Recently, such detergents have also begun to have an impact on the European and Japanese markets (Fig. 60).

Table 25. Frame formulations for powdered heavy-duty detergents [104]

Ingredients	Examples	Composition, %					
		Western Europe		Japan		United States	
		with phosphate	without phosphate	with phosphate	without phosphate	with phosphate	without phosphate
Anionic surfactants	alkylbenzenesulfonates	5–10	5–10	5–15	5–15	0–15	0–20
	fatty alcohol sulfates	1–3		0–10	0–10		
	fatty alcohol ether sulfates					0–12	0–10
	α-olefinsulfonates			0–15	0–15		
Nonionic surfactants	alkyl poly(ethylene glycol) ethers, nonylphenyl poly(ethylene glycol) ethers	3–11	3–6	0–2	0–2	0–17	0–17
Suds-controlling agents	soaps, silicon oils, paraffins	0.1–3.5	0.1–3.5	1–3	1–3	0–1.0	0–0.6
Foam boosters	fatty acid monoethanolamides	0–2					
Chelating agents	sodium triphosphate	20–40		10–20		23–55	
Ion exchangers	zeolite 4 A, poly(acrylic acids)	2–20	20–30	0–2	10–20		0–45
Alkalies	sodium carbonate	0–15	5–10	5–20	5–20	3–22	10–35
Cobuilders	sodium nitrilotriacetate, sodium citrate	0–4	0–4				
Bleaching agents	sodium perborate, sodium percarbonate	10–25	20–25	0–5	0–5	0–5	0–5
Bleach activators	tetraacetylethylenediamine	0–5	0–2				
Bleach stabilizers	ethylenediaminetetraacetate, phosphonates	0.2–0.5	0.2–0.5				
Fabric softeners	quats, clays				0–5	0–5	0–5
Antiredeposition agents	cellulose ethers	0.5–1.5	0.5–1.5	0–2	0–2	0–0.5	0–0.5
Enzymes	proteases, amylases	0.3–0.8	0.3–0.8	0–0.5	0–0.5	0–2.5	0–2.5
Optical brighteners	stilbenedisulfonic acid, bis(styryl)biphenyl derivatives	0.1–0.3	0.1–0.3	0.1–0.8	0.1–0.8	0.05–0.25	0.05–0.25
Anticorrosion agents	sodium silicate	2–6	2–6	5–15	5–15	1–10	0–25
Fragrances		+	+	+	+	+	+
Dyes and blueing agents		+	+	+	+	+	+
Formulation aids						0–1.0	0–1.0
Fillers and water	sodium sulfate	balance	balance	balance	balance	balance	balance

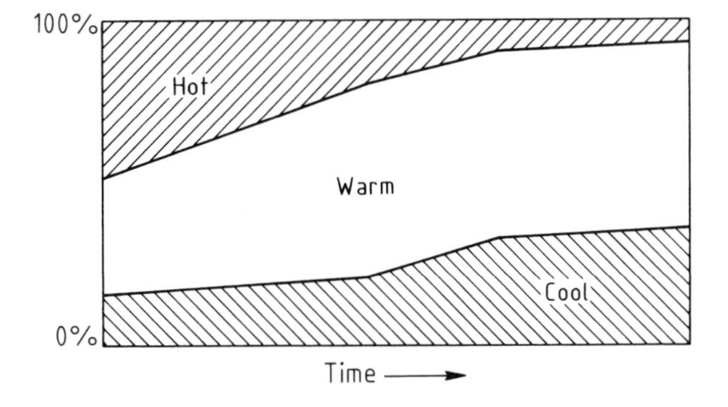

Figure 59. Home laundry washing machine temperature trends in the United States [146]

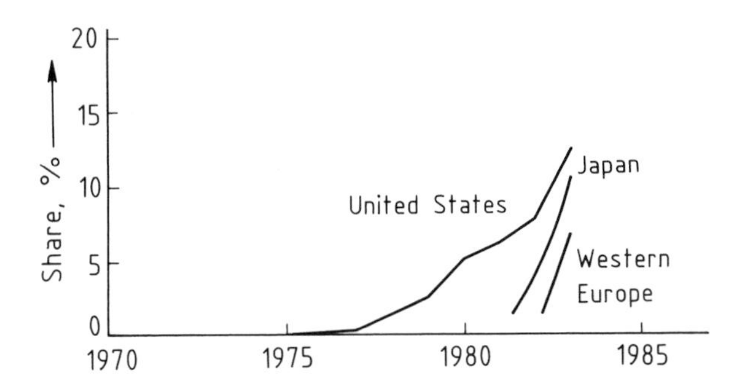

Figure 60. Market share of heavy-duty detergents containing fabric softeners [42]

4.1.2 Liquid Heavy-Duty Detergents

Liquid detergents have long been common in the United States for a variety of reasons, and the market trend continues to be upward. Until recently, such products have been insignificant in Europe [147], although liquid heavy-duty detergents have been available on the European market since 1981. These products are distinctive because of their relatively high surfactant content (up to ca. 40 %). For reasons of solubility and stability, they seldom contain builders and generally are devoid of bleaching agents. Their effectiveness is concentrated on the removal of grease and greasy soil, especially at wash temperatures < 60 °C and at doses of 6–12 g/L (depending on water hardness and the amount of soil to be removed).

Currently, liquid heavy-duty detergents represent < 10 % of the market in both Western Europe and Japan. In the United States, their market share is already > 30 %, and products are offered both with and without added builders (Fig. 61). Formulation ranges for liquid heavy-duty detergents with and without builders in Europe, the United States, and Japan are given in Table 26. Liquid detergents with builders in the United States typically contain 15–30 % surfactants, but the overall amount of surfactants in products without builders can be as high as 55 %. There are

Table 26. Frame formulations for liquid heavy-duty detergents [104]

Ingredients	Examples	Composition, %					
		Western Europe		Japan		United States	
		with builders	without builders	with builders	without builders	with builders	without builders
Anionic surfactants	alkylbenzenesulfonates	5–7	10–15	5–15		5–17	0–10
	soaps		10–15	10–20		0–14	
	fatty alcohol ether sulfates			5–10	15–25	0–15	0–12
Nonionic surfactants	alkyl poly(ethylene glycol) ethers	2–5	10–15	4–10	10–35	5–11	15–35
Suds-controlling agents	soaps	1–2	3–5				
Foam boosters	fatty acid alkanolamides	0–2					
Enzymes	proteases	0.3–0.5	0.6–0.8	0.1–0.5	0.2–0.8	0–1.6	0–2.3
Builders	potassium diphosphate, sodium triphosphate	20–25					
	sodium citrate, sodium silicate		0–3	3–7		6–12	
Formulation aids	xylenesulfonates, ethanol, propylene glycol	3–6	6–12	10–15	5–15	7–14	5–12
Optical brighteners	stilbenedisulfonic acid, bis(styryl)biphenyl derivatives	0.15–0.25	0.15–0.25	0.1–0.3	0.1–0.3	0.1–0.25	0.1–0.25
Stabilizers	triethanolamine, chelating agents		1–3	1–3	1–5		
Fabric softeners	quats, clays					0–2	0
Fragrances		+	+	+	+	+	+
Dyes		+	+	+	+	+	+
Water		balance	balance	balance	balance	balance	balance

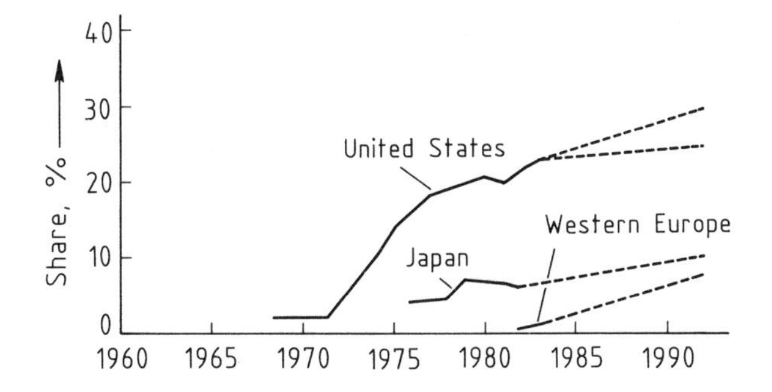

Figure 61. Market share of heavy-duty liquid detergents [42]

several reasons for the importance of liquid detergents in the United States, including laws regulating phosphate content, typical modes of washing machine design and operation, and washing customs. In particular, the tightening of phosphate regulations led to an enormous increase in the sale of builder-free detergents. Another factor is the lack of a tradition in the United States for subjecting laundry to a presoak and the fact that detergent is normally added directly rather than through a dispenser. The usual bleaching agent in the United States is hypochlorite, which is introduced separately from the detergent.

4.2 Specialty Detergents

Specialty detergents play a relatively minor role in the United States, but they are quite important in Western Europe.

Specialty detergents are products developed for use with specific types of household fabrics. Such detergents are generally employed with washing machines and usually require the use of special cycles (e.g., a wool cycle to prevent felting, a colored wash cycle to eliminate dye transfer, or a curtain cycle to prevent wrinkling).

4.2.1 Powdered Specialty Detergents

Distinction is made in Western Europe among products of the following types:

detergents for easy-care and colored fabrics
detergents for wool
detergents for curtains
detergents for manual wash

Table 27. General formulations of West European powdered specialty detergents

Ingredients	Composition, %			
	Detergents for delicate and colored laundry	Detergents for woolens	Detergents for curtains	Manual washing detergents
Anionic surfactants (alkylbenzenesulfonates, fatty alcohol ether sulfates)	5–15	0–15	0–10	12–25
Nonionic surfactants (fatty alcohol poly(ethylene glycol) ethers)	1–5	2.0–25*	2–7	1–4
Soaps	1–5	0–5	1–4	0–5
Cationic surfactants (dialkyldimethylammonium chloride)		0–5**		
Sodium triphosphate	25–40	25–35	25–40	25–35
Sodium perborate			0–12	
Sodium silicate	2–7	2–7	3–7	3–9
Antiredeposition agents	0.5–1.5	0.5–1.5	0.5–1.5	0.5–1.5
Enzymes	0–0.4			0.2–0.5
Optical brighteners	0–0.2		0.1–0.2	0–0.1
Fragrances	+	+	+	+
Fillers and water	balance	balance	balance	balance

* Higher amounts of nonionic surfactants are contained in liquid detergents only. ** Cationic surfactants in liquid detergents for woolens only.

Many of these products are appropriate for either machine or manual washing. Most are free of sodium perborate, and detergents for wool lack fluorescent whiteners. Detergents specifically designed for easy-care and colored fabrics contain neither sodium perborate nor fluorescent whitening agents because of the sensitivity of such fabrics. Detergents in this category are particularly useful with dyes that may be subject to oxidation or with pastel fabrics that might experience color shifts if treated with fluorescent whitening agents.

Detergents for wool are primarily intended for use in a washing machine. Particular care must be exercised to prevent damage to the sensitive fibers that constitute natural wool, including maintenance of a low temperature and avoidance of vigorous mechanical action.

Detergents specially formulated for curtains usually contain especially effective soil antiredeposition agents.

Heavily foaming detergents for manual washing are intended for occasions when a small amount of laundry is done in the sink.

Table 27 provides formulation ranges for these various types of specialty detergent.

4.2.2 Liquid Specialty Detergents

Liquid specialty detergents have long been on the market even in Western Europe, which is in contrast to the availability of liquid heavy-duty detergents. Many liquid detergents are intended for manual use; however, specialty liquid products also exist that are intended for machine washing application, including detergents for wool and detergents for curtains. Liquid detergents for wool may be free of anionic surfactants, in which case they usually contain mixtures of cationic and nonionic materials. In such a case, the cationic agents act as fabric softeners to help keep wool soft and fluffy. Table 28 describes typical formulations for liquid specialty detergents.

Table 28. General formulations of West European liquid specialty detergents [147]

Ingredients	Composition, %		
	Detergents for woolens		Detergents for curtains
	with incorporated softeners	without incorporated softeners	
Anionic surfactants (alkylbenzenesulfonates, fatty alcohol ethylene glycol ether sulfates)		10–30	0–8
Nonionic surfactants (fatty alcohol poly(ethylene glycol) ethers, fatty acid amides)	20–30	2–5	15–30
Cationic surfactants (dialkyldimethylammonium chloride)	1–5		
Ethanol, propylene glycol	0–10	0–10	0–5
Toluenesulfonates, xylenesulfonates, cumenesulfonates		0–3	
Builders (potassium diphosphate, sodium citrate)		0–15	2–5
Optical brighteners			+
Fragrances, dyes	+	+	+
Water	60–70	60–80	65–75

4.3 Laundry Aids

Laundry aids are products developed to meet the varying needs of often widely divergent washing techniques and practices in different parts of the world. For example, most powdered heavy-duty detergents in Western Europe contain bleaching agents; those in the United States and Japan do not, and thus, separate addition of bleach is required. Depending on their use, laundry aids can be divided into two categories: pretreatment aids and boosters.

4.3.1 Pretreatment Aids

The most important pretreatment aids are laundry water softeners, soaking agents, and soil and stain removers.

Laundry water softeners exist both in the United States and in Western Europe. European manufacturers recommend that these softeners be used in combination with detergents for any application involving water of medium to high hardness. In the Federal Republic of Germany, for example, laws effectively require that in such cases detergent amounts be based on soft water dosage. European laundry softeners generally contain sequestering agents of the oligomeric phosphate or nitrilotriacetate type, along with ion exchangers, usually zeolite 4 A and polymeric carboxylic acids. In the United States and Japan, sodium triphosphate and sodium carbonate dominate (Table 29).

Soaking agents as found in Europe are washing aids usually adjusted to be strongly alkaline (pH 12–12.5) so that they will loosen particularly stubborn and firmly adhering soil (Table 30). Swelling phenomena play a major role in their effectiveness.

Soil and stain removers are normally products high in surfactant content. Their direct application aids in the removal of localized greasy soil (Table 31). Such materials are applied to soiled areas immediately prior to a standard washing process. They are supplied either in paste or spray form. The major targets of such products are fabrics of the easy-care type, which require low-temperature washing. The regular use of soil and stain removers can also prevent the buildup of stains on problem areas such as shirt collars and sleeves.

Spray stain removers are comprised largely of mixtures of solvents and surfactants. Solvents similar to those employed in dry cleaning loosen grease so that it can be emulsified by appropriate surfactant mixtures. Sprays generally work more rapidly than paste formulations. Proper application of such products keeps undesirable

Table 29. Formulations of laundry water softeners [148]

Ingredients	Composition, %	
	Europe	United States
Sodium triphosphate	20–50	50–60
Sodium nitrilotriacetate	10–20	
Zeolite 4 A	20–40	
Polycarboxylic acids	1–5	
Sodium carbonate		15–25
Others (sulfates, etc.)	balance	balance

Table 30. Formulation of European presoaking products [148]

Ingredients	Composition, %
Alkylbenzenesulfonates	2–7
Alcohol polyglycol ethers	0–2
Soaps	0–2
Sodium carbonate	50–80
Sodium silicate	5–10
Carboxymethyl cellulose	0–2
Dyes, fragrances	+
Water	balance

Table 31. Formulations of prewash soil and stain removers in Europe, United States, and Japan [148]

Ingredients	Composition, %					
	Europe		United States[a]		Japan	
	Paste	Spray	Aerosol	Non-aerosol	Liquid	Aerosol
Anionic surfactants (alkylbenzenesulfonates, alcohol sulfates, alcohol ether sulfates)	15–30				5–20	
Nonionic surfactants (alcohol polyglycol ethers, fatty acid ethanolamides, fatty acid esters)	3–10	20–40[b]	15–30[b]	5–10[b]	5–15	20–30[b]
Hydrocarbons		20–45	20–70			50–70
Methylene chloride		20–35				
Xylenesulfonate				0–5		
Propellants (CO_2, butane, propane)		1–4	10–15			10–15
Dyes, fragrances, FWA[c], water	balance	balance	balance	balance	balance	balance

[a] Prewash paste not on the market. [b] Alcohol polyglycol ethers. [c] Fluorescent whitening agents.

rings from forming around stains, a consequence that is virtually unavoidable if spots are subjected to local treatment with pure solvents instead.

4.3.2 Boosters

Boosters are products added separately to detergents to exert specific influences on the washing process and thereby improve its effectiveness. The principal types of boosters offered on the market are bleaching agents and laundry boosters.

Table 32. Formulations of dry all-fabric bleaches in the United States and Japan [148]

Ingredients	Composition, %	
	United States	Japan
Sodium perborate	10–30	
Sodium percarbonate		70–85
Sodium triphosphate, sodium carbonate, sodium sulfate	40–85	15–30
Fluorescent whitening agents	0–0.8	
Enzymes	0–2	0–1
Dyes, fragrances	+	+

Table 33. Formulation of laundry boosters in the United States [148]

Ingredients	Composition, %
Sodium carbonate	20–25
Sodium silicate	5–7
Sodium sulfate, sodium chloride	40–50
Alkylbenzenesulfonates	6–10
Enzymes	0.5–1
Water	balance

Bleaching agents are particularly common in the United States and Japan [149], where they are available in both powdered and liquid form. Powdered bleaching agents generally contain sodium perborate or sodium percarbonate (Table 32), whereas most liquid bleaching agents are 5–6 % solutions of sodium hypochlorite.

Laundry boosters are marketed largely in the United States; various formulations exist. They contain either sodium triphosphate, sodium citrate, or sodium carbonate, usually in combination with surfactants. Enzymes are often present as well (Table 33). "Laundry booster sheets," a special form of laundry booster, are sold in Western Europe. These consist of pieces of fabric that have been impregnated with special surfactants and bleach activators. They are intended for one-time use together with a detergent and are added directly to the laundry. The cloth is discarded after the washing process has been completed. Boosters of this kind are said to improve wash effectiveness. Their manufacturers claim that the effect is particularly noticeable with greasy soil and at wash temperatures below 60 °C.

4.4 Aftertreatment Aids [150], [151]

After the washing process has been completed and soil removal has been largely accomplished, fabrics are sometimes subjected to some type of aftertreatment. The goal is to increase the usefulness of laundry by restoring characteristics that have suffered in the course of the wash. Needs in this respect can vary considerably, depending on the fabric involved. Thus, products in this category may be called on to provide elastic stiffness, improved fit, and body (shirts and blouses); smoothness, sheen, and rigidity (tablecloths and napkins); good drop (curtains); fluffiness and softness (undergarments, towels, and bath robes); or antistatic properties (easy-care articles made from synthetic fibers). To achieve such effects, the following product groups are marketed:

fabric softeners
starches and stiffeners
fabric formers
laundry dryer aids

4.4.1 Fabric Softeners

Textiles washed by machine are subjected to greater mechanical stress than those washed by hand. Indeed, machine-washed laundry is so severely pummeled that the pile of the fibers at the fabric surface is reduced to an extreme state of disarray, especially in the case of natural fibers. During subsequent drying in relatively static air (e.g., when laundry is dried indoors), this condition tends to become fixed into the fabrics and the laundry acquires a harsh feel. Addition of a fabric softener in the final rinse (rinse cycle softener) results in fabrics that feel softer.

The conditions described above are found mainly in Europe and Japan. Fabric softeners play a lesser role in the United States, where most laundry is dried in a mechanical dryer and the tumbling that accompanies the process accomplishes its own softening effect. Thus, the chief task of household fabric softeners in the United States is to confer antistatic properties.

The principal active ingredients in commercially available rinse cycle softeners are usually cationic surfactants of the quaternary ammonium type [152], [153].

Recently, the European market has witnessed some success on the part of concentrated fabric softeners, which compete alongside the more conventional formulations (characterized in Table 34). The former are normally 3-fold (15% active ingredient) or 10-fold (50% active ingredient) concentrates, whereas the usual brands contain ca. 5% active ingredients. Significant differences exist in the use of the term "concentrate" in Europe and in the United States. Fabric softeners in the United

Table 34. Formulations of fabric softeners in Europe, the United States, and Japan

Ingredients	Composition, %		
	Europe	United States	Japan
Dialkyldimethylammonium chloride	1–9	4–15	4–5
Alkylimidazolinium methyl sulfate	40–50*	2–15	
Methylbis(alkylamidoethyl)-2-hydroxyethylammonium methyl sulfate		2–10	
Alcohol polyglycol ethers, nonylphenol polyglycol ethers	0–3		0–3
Fluorescent whitening agents	0–0.2	0–0.2	
Preservatives	0.1–0.5	0.1–0.5	0–0.5
Alkylbenzyldimethylammonium chloride	0–1.5		
Dyes, fragrances	+	+	+
Water	balance	balance	balance

* Only in 10-fold concentrated products.

Table 35. Sorption of distearyldimethylammonium chloride* [50]

Textiles	Adsorbed amount**			
	mg/g	mol/g	mg/m^2	%
Wool	1.20	2.06	301	100
Resin-finished cotton	1.18	2.01	133	98
Cotton	1.17	2.00	169	96
Polyester/cotton	1.17	2.00	110	98
Polyamide	0.96	1.63	66	79
Polyacrylonitrile	0.90	1.53	68	74
Polyester	0.57	0.97	109	47

* Equilibrium conditions: time = 60 min, t = 23 °C, bath ratio = 1:10; initial concentration: 120 mg/L.

** Based on the mass and geometric surface of the textiles.

States generally contain 3–4 % active material, and products with as little as 7–8 % active material are described as concentrates [151].

The formulation of products with higher amounts of active material requires a well-balanced system of selected emulsifiers if good dispersion stability is to be maintained. In addition to the classical ammonium and imidazolinium compounds, increasing numbers of other cationic surfactants with antistatic and fabric softening properties have been described in the literature on fabric softener concentrates, including polyammonium, bisimidazolinium, and alkylpyridinium derivatives and certain cationic polymers.

To prevent disruptive interactions between the anionic surfactants of a detergent and the cationic surfactants of a fabric softener, the latter must be introduced only in the rinse cycle.

When employed in appropriate concentrations, cationic surfactants are attracted nearly quantitatively to natural fibers, in contrast to their behavior with various synthetic fibers (Table 35) [154]–[156].

Large excesses of fabric softener are to be avoided; otherwise, the absorbancy of the fabric suffers, a consequence that is particularly serious with towels.

Fabric softeners have the additional benefit that they confer antistatic properties on fabrics. This prevents the buildup of electrostatic charge on synthetic fibers, which in turn eliminates such disagreeable characteristics as fabric cling, crackling noises, and dust attraction.

4.4.2 Starches and Stiffeners

If stiffness and body are desired rather than soft and fluffy laundry, wash stiffeners can be added to the process. The usual agents for this purpose include natural starch derived from rice, corn, or potato, which can be used to produce extremely stiff fabric. However, synthetic polymeric stiffeners have become increasingly important recently. These can contribute a more modest degree of stiffness, which is more closely attuned to contemporary taste. Products of this type are generally liquid, which are easier to apply than natural starch. Stiffeners are supplied as dispersions and contain, in addition to a small amount of starch, substances such as poly(vinyl acetate), which has been partially hydrolyzed to poly(vinyl alcohol). Table 36 shows a formulation for typical European liquid stiffening agents. Such products are offered not only as dispersions, but also as aerosols and sprays.

Such stiffening agents are called permanent stiffeners because, unlike starch, their effectiveness endures through several wash cycles. This property also has its negative aspects, however, since poly(vinyl acetate) film on a fabric can attract both soil and dyes, thereby leading to discoloration.

Table 36. Formulation of liquid stiffeners in Europe

Ingredients	Composition, %
Alkyl/alkylaryl polyglycol ethers	0.1–2
Poly(vinyl acetates) (partially saponified)	15–40
Starches	0–5
Poly(ethylene glycols)	0.5–1.5
Fluorescent whitening agents	0.01–0.3
Dyes	0.1–0.4
Water	balance

4.4.3 Fabric Formers

Fabric formers are encountered mainly in Western Europe and offer certain advantages over starch and stiffeners with respect to ease of application, effectiveness, and secondary properties that are conferred. Textiles that are treated with a fabric former become firm, but are not cast into a stiff layer. Such products are liquids that are based on copolymers of vinyl acetate with unsaturated organic acids. They are supplied in ready-to-use form, which unlike stiffeners can be conveniently added to laundry in a washing machine. Polymeric wax additives confer the benefit of simplified ironing and provide additional smoothness to the fabric surface. Table 37 provides information about characteristic formulation ranges.

In contrast to stiffeners, fabric formers leave no deposits that might attract soil and pigments from the wash liquor. The active ingredients are soluble in a weakly alkaline medium and, thus, are readily washed off.

Fabric formers also exert a soil-release effect; i.e., they simplify soil removal in subsequent wash operations.

Table 37. Formulation of fabric formers in Europe

Ingredients	Composition, %
Alkyl sulfates, alcohol polyglycol ethers	0–2
Foam inhibitors	0.1–2
Copolymers (vinyl acetate/acid)	20–45
Polywaxes	0.5–2.5
Fluorescent whitening agents	0.01–0.3
Fragrances	0.1–0.4
Preservatives	0.3
Water	balance

4.4.4 Laundry Dryer Aids [157]–[159]

Laundry dryers ("tumblers") are far more widespread in the United States than in Europe. By 1984, ca. 66% of all American households were equipped with dryers, as compared to only ca. 10% of the households in Western Europe. For this reason, laundry dryer aids have held a significant place in the United States market since the early 1970s. Products in this category are introduced into the dryer along with damp, spin-dried laundry. During the drying process, they provide the laundry with both a measure of softness and a pleasant aroma. Most importantly, however, they prevent static buildup on the fabric. The latter point is particularly significant in the United States, where synthetic fabrics are widely used. Products of this type have also begun to appear on the European market.

Laundry dryer aids can be placed in one of four classes depending on their mode of application:

1) Aerosol foam, which is first sprayed on a cloth or some laundry item and then added to the moist laundry in the dryer
2) Aerosol spray for coating the inside of the empty dryer drum prior to addition of the laundry
3) Pads saturated with active material, which are attached by means of an adhesive patch to one of the paddles in the dryer drum
4) Sheets, which serve as carrier, made of polyurethane foam or some nonwoven material and impregnated with both fabric softeners and temperature-resistant fragrance oils

During the drying process, active ingredients from the dryer aid are transferred to fabrics by frictional contact. The sheet materials are designed for single use and are discarded at the end of the drying cycle. Only carrier sheets — made from polyurethane foam or especially from cellulose or polyester nonwovens — have proved to be market successes.

The preferred impregnating agent for carrier sheets is distearyldimethylammonium methyl sulfate rather than the corresponding chloride since the latter can cause corrosion in the dryer. An additional reason for the popularity of dryer aids in the United States is the fact that washing machines in the United States generally lack dispensers for separate introduction of fabric softener.

5 Institutional Detergents

Even though the principles that determine the effectiveness of detergents for household and commercial laundries are the same, detergents for large-scale institutional use generally differ to the extent that they must be designed to meet the special circumstances associated with laundry on an industrial scale [160], [161]. Commercial laundries, in contrast to those in the household, normally have soft water at their disposal, usually obtained from softening systems that employ softening filters. The base of the filter holder in such a device contains outlets for the water and is surmounted by several gravel layers differing in particle size. Above these is located the actual "filtering" material, an ion exchanger. Many types of ion-exchange resins have been successfully employed.

Since high-quality water is no longer available in unlimited quantity, water treatment has become increasingly expensive. This in turn has led cost-conscious laundries not only to seek processes that conserve water, but also to find ways by which water can be recycled.

The development of continuous-process commercial laundry equipment with efficient water and energy use and a turnover rate of > 500 kg per hour and unit has necessitated the concurrent development of appropriate detergent combinations. A further impetus has been the desire by commercial laundries to modify their processes to deal more effectively with particular types of laundry and soil or to enhance certain aspects of the cleansing action. In some cases, the answer has been to employ *partially built* detergent mixtures, in which one component (e.g., soap or synthetic surfactants, di- or triphosphate, alkali, or perborate) is dominant and others may be absent altogether. Proper use of such a partially built detergent requires expertise and a highly reliable personnel. More common are combinations that differ from completely built detergents only by the absence of a bleaching agent. This absence permits the individual launderer to select a particular type of bleach appropriate to a specific batch of laundry or to refrain from bleaching entirely. In contrast to the household detergent market the commercial sector employs not only sodium perborate and hydrogen peroxide solutions as bleaching agents, but also alkaline hypochlorite and organic chlorine carriers.

Most recent developments in Western Europe have been toward simplified utilization of a single type of product, analogous to the heavy-duty detergents for household use, supplemented perhaps by a partially built product employed in a pretreatment step.

5.1 Partially Built Detergents

Until the late 1950s standard commercial washing machines in Western Europe employed multiple wash cycles (two- or three-shot systems). For such machines, nearly the only detergents available were the so-called partially built detergents, which came in various forms tailored to specific applications. Thus, there were alkaline formulations comprised of soda, metasilicate, and orthophosphate; specialty surfactants in the form of soap or grease-removing synthetic surfactant pastes; mixtures of condensed phosphates; and, finally, liquid or powdered bleaching agents. All of these were later refined by the addition of fluorescent whitening agents, soil antiredeposition agents, and magnesium silicate, the latter acting as an oxygen bleach stabilizer.

The availability of such specialized partially built products reduced the effort required for laundering from a detergent standpoint, but laundry processes as a whole still remained inconvenient and complex, and their proper operation required skill and experience.

In the meantime, partially built products have lost much of their former market strength, especially in the Federal Republic of Germany. Currently, these products are utilized only as wash liquor additives in special situations in which unusual conditions demand them, e.g., extremely high soil levels or inconvenient water conditions. In fact, these are the only reasons for such products to continue to be marketed (Table 38).

In the United States, partially built detergents as well as raw materials and simple raw material mixtures are all marketed as soaps, builders, and alkalies. The so-called soaps are not always pure soaps in the chemical sense of the word, but rather are often mixtures of soaps and synthetic detergents. Alkalies and builders usually consist of soda, metasilicate, and sequestering agents and are always characterized by high alkalinity.

In Japan, laundries commonly use only soap powders and metasilicate. Products containing enzymes are occasionally employed for added washing power. The preferred surfactants in partially and completely built detergents in Japan are alkane- and olefinsulfonates.

5.2 Detergents for the Pre- and Main Wash

The extremely labor-intensive and time-consuming multiple wash cycle methods were reduced early in 1960 in Western Europe to two stages as a result of the introduction of two detergents, one for the prewash and one for the main wash,

Table 38. Formulations* of various types of detergents for institutional use

Components	Detergents				Partially built products		
	Perborate-containing	Completely built	Main wash	Prewash	Alkalies	Surfactants	Bleaching agents
Surfactants	× ×	× ×	× × ×	×	(×)	× × × ×	
Sodium triphosphate	×(× × ×)	×(× × ×)	×	× × ×			
Alkalies (soda, metasilicate, orthophosphate)	×(× ×)	×(× ×)	×	× × ×	× × × ×		
Bleaching agents	×		×				× ×
Fluorescent whitening agents	×	×	×	(×)	(×)	(×)	(×)
Additives (CMC, $MgSiO_3$)	×	×	×	(×)	(×)	(×)	(×)
Enzymes	(×)			(×)			

* × – × × × × = varying extent of individual components; ( ) = component may be present.

carefully matched in their compositions. The new process, called the *duplex technique*, permitted washing with a standard drum-type institutional washing machine in a total average time of 45–50 min inclusive of rinsing. One prerequisite to achieving such rapid results, however, was the availability of a steam ratio adequate to allow rapid heating. The prewash detergent was a direct descendent of partially built detergents of the alkali type. In addition to an alkali framework, high-quality prewash detergents contain a small amount of surfactants to improve wetting and wash effectiveness, as well as generally large amounts of sequestering phosphates (see Table 38).

In addition, main wash detergents usually include a certain amount of perborate, obviating the need to separately introduce bleach. Perborate-containing main wash detergents can be compared in their composition to heavy-duty detergents. The term "main wash detergents" in the institutional sector usually indicates detergents intended for use in combination with prewash detergents in a two-cycle process. The increasing popularity of continuous batch washing machines with counterflow systems has led to a significant drop in worldwide market share for classical pre- and main wash detergents.

5.3 Perborate-Containing Detergents

Detergents for institutional use have also developed toward products mentioned previously that are comparable to heavy-duty detergents in the household sector. Institutional perborate-containing detergents are significant only in Western Europe

Table 39. Formulation of a perborate-containing detergent for institutional use

Components	Examples	Composition, %
Surfactants	alkylbenzenesulfonates, soaps, fatty alcohol polyglycol ethers	10–15
Sequestering agents	pentasodium triphosphate	20–30
Alkalies	soda and silicates	10–20
Bleaching agents	sodium perborate	20–30
Fluorescent whitening agents	stilbene and pyrazoline derivatives	0.1–0.3
Antiredeposition agents	carboxymethyl cellulose	0.5–2
Corrosion inhibitors	water glass	3–5
Foam inhibitors	docosanoic acid	2–3
Stabilizers	ethylenediaminetetraacetate, magnesium silicate	0.2–2
Fragrance oils		0.1–0.2
Coloring agents		0.0007–0.001

and in other regions where very hot water is used for laundering. In Japan, percarbonate is preferred over perborate.

The development of perborate-containing institutional detergents was strongly influenced by the introduction of counterflow equipment in the 1950s. As these systems were increasingly automated, addition of individual partially built detergents as separate components became less practical. The formulation of a typical perborate-containing detergent for institutional use is shown in Table 39.

5.4 Completely Built Detergents

A variant among the institutional detergent types is the completely built detergent. In principle, such products are constructed like household heavy-duty detergents, but they lack perborate (cf. Table 38). Currently, they are used quite commonly in continuous batch washing machines with counterflow systems. The launderer must decide whether an oxygen or a chlorine bleach is to be added, which also permits the option of omitting bleach altogether in the interest of treating the laundry especially gently.

This product group includes a very large number of detergents designated in many different ways worldwide. For example, the majority of the detergents marketed in the United States belong to this group. These are often distinguished from partially built products by calling them one-shot detergents.

5.5 Specialty Detergents

Specialty detergents are products formulated in such a way as to meet the demands of particular kinds of laundry or particular laundering processes. These include, for example, detergents without whitening agents for easy-care and colored fabrics, detergents for work clothes, and enzyme-containing products for soils especially rich in protein. Also included are disinfectant additives for hospital use; such detergents are specifically designed to handle laundry originating in isolation wards and, therefore, requiring thermal and chemothermal sterilization. Common disinfectants for this purpose include chlorinated phenols, 2-phenylphenolates, sodium *N*-chloro-*p*-toluenesulfonamide (chloramine-T), dichlorodimethylhydantoin, and other organic chlorine carriers which liberate hypochlorous acid in the presence of water. They have acquired increasing importance with the advent of centralized laundries and the arrangements necessary to prevent the spread of infections. Disinfecting procedures of this sort are most common in Western Europe, but absent in Japan. Laundry sterilization in the United States is accomplished either thermally or by active chlorine bleaches.

6 Production of Powdered Detergents

6.1 Requirements

Detergents must be viewed as consumer goods derived from industrial production operations. This means, for example, that their precise formulation is determined by more than just their intended purpose: the number and kind of raw materials to be processed has a direct effect on production and can, therefore, influence aspects of the formulation as well. Production technology is responsible for establishing certain important qualitative characteristics of a product:

degree of retention of raw material activity
moisture content
bulk density
homogeneity
particle size distribution
flowability
dust behavior
dispersing and flushing properties

Until the late 1960s, most changes in detergent make-up originated primarily from factors associated with product application or with developments in the raw material sector. For the last two decades, however, developments have been influenced to an increasing degree by economic constraints and by demands imposed for the protection of both consumers and the environment [162], [163]. Therefore, any proposal for instituting a new detergent production facility must be evaluated from the following perspectives:

1) Cost of investment, processing, materials, and energy
2) Flexibility with respect to possible raw materials and potential consumer products
3) Legal standards taking effect for plant operations, environmental impact, and product characteristics

6.2 Production Processes

6.2.1 Engineering Principles and Historical Review

Some of the surfactants, builders, and additives used in detergents are supplied as aqueous solutions or suspensions. Therefore, one of the fundamental goals of all production processes is to eliminate this water. Generally, the available methods are (1) crystallization techniques, in which the water is bound by substances capable of forming hydrates (e.g., soda, silicates, phosphates, and either sodium sulfate or sodium borate); and (2) evaporative methods, whereby excess water is removed thermally.

In the early days of detergent manufacture, only crystallization methods were employed. Soap paste, soda, and water glass solution were simply mixed and the mixture spread out in a storage shed for crystallization. The solidified mass was then pulverized mechanically and packaged for shipment.

KRAUSE in 1912 [164] first succeeded in applying the spray-drying method [165] on an industrial scale, although the principle itself was already known. In this approach, the mixture of detergent components was transferred to a spray tower and there distributed by attached spray arms onto a disk rotating at 3000–15 000 min^{-1}. Simultaneously, a stream of cold air was passed into the tower. Centrifugal force caused the liquid droplets to be dispersed as a fog in which crystallization then occurred rapidly. The complete crystallization succeeded in cooling towers coupled to the spray tower.

A serious shortcoming of this simple crystallization technique was the severe limitations imposed on the potential nature of the detergent formulation. The extent to which liquid components could be tolerated was limited by the drying capacity of the solids present. Consequently, the process was later modified to employ hot spray techniques, in which hot air was introduced into the spray tower. However, all washing powders produced by rotary disk atomization still shared one disadvantage: both their dust content and their flowability tended to be unsatisfactory. This difficulty was solved by the conversion to "bead form" detergents, the production of which was based on developments pioneered in the United States [166], [167]. A mixture of solid and liquid raw materials, called a slurry, is sprayed under pressure directly through a set of nozzles located at the top of a tower. The resulting droplets are larger than those produced by the rotary disk method. When these drops contact hot air, they are essentially blown, creating hollow spheres known as beads (see Fig. 62).

The powder characteristics associated with material prepared by this method are subject to a considerable degree of control, and ca. 80% of all detergents are estimated to be made in this way currently. Therefore, this method is examined in more detail in the sections that follow.

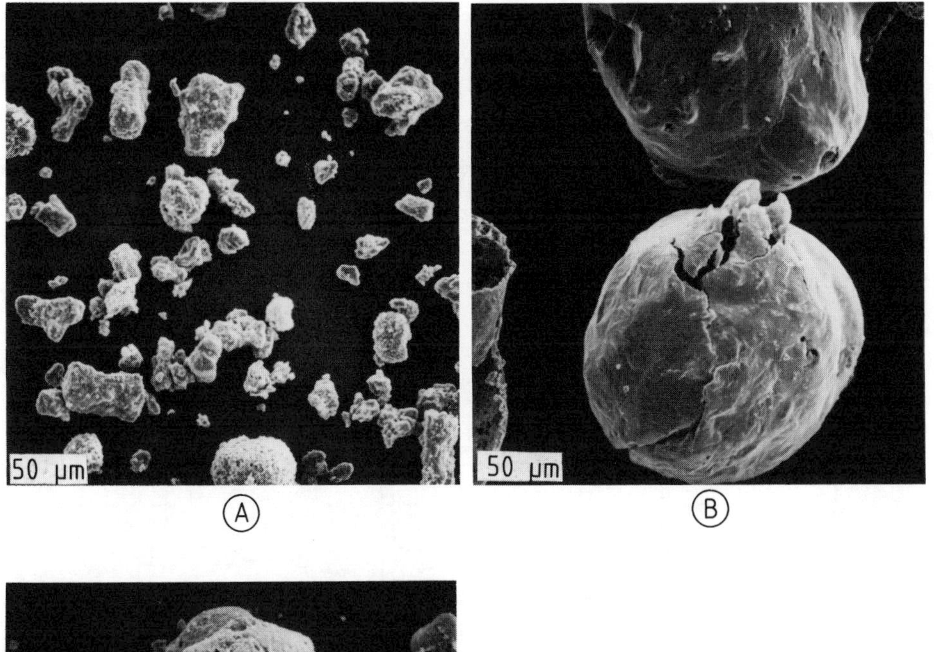

Figure 62. Scanning electron micrographs of tower powders
Product prepared by: A) Crystallization drying in a rotary disk atomizing tower; B) Cocurrent high-pressure nozzle atomization; C) Countercurrent high-pressure nozzle atomization

6.2.2 High-Pressure Nozzle Atomization [168]–[172]

Materials Flow. The first step in this process is preparation of a slurry from the thermally stable components, i.e., those that neither decompose nor vaporize under the influence of hot air in the tower. Individual solid and liquid components are

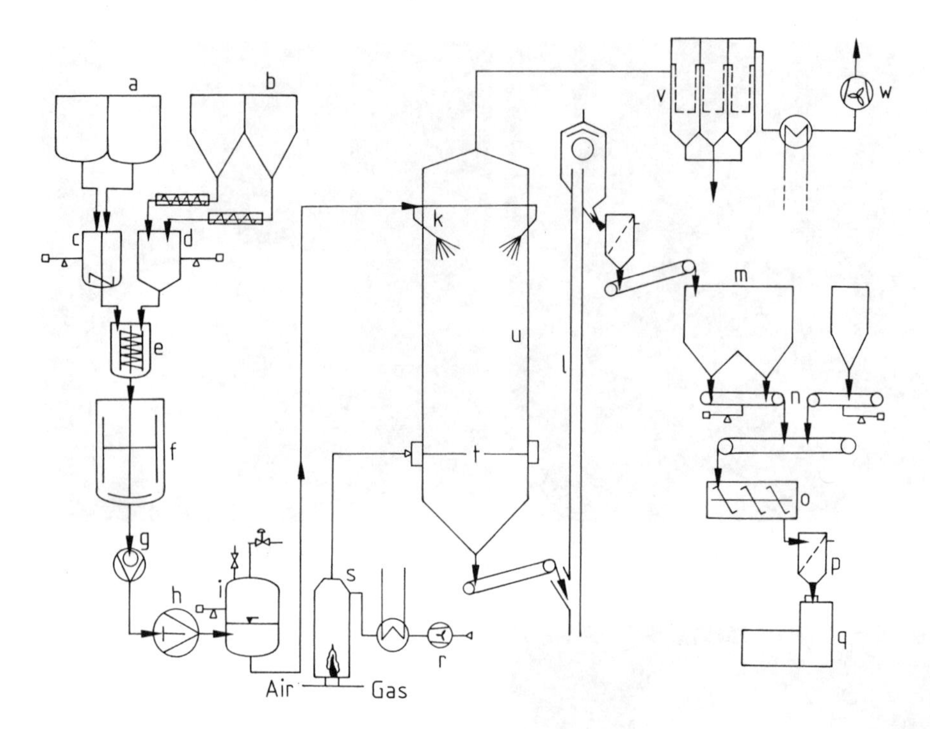

Figure 63. Flow diagram of a spray-drying process for detergents (countercurrent high-pressure nozzle atomization)

a) Storage tanks for liquid raw materials; b) Storage silos for solid raw materials; c) Liquids weighing vessel; d) Solids weighing vessel; e) Mixing vessel; f) Intermediate tank; g) Booster pump; h) High-pressure pump; i) Air vessel; k) Nozzles; l) Airlift; m) Storage bunker; n) Belt conveyor scales; o) Powder mixer; p) Sieve; q) Packaging machine; r) Air inlet fan; s) Burner; t) Ring channel; u) Spray tower; v) Bag filter; w) Exhaust fan

drawn off from silos or tanks and introduced batchwise into scales (Fig. 63). Water is added only as required to maintain a manageable viscosity. Concentration data reported for such a slurry are based on percentage of either dry material or of dried tower powder (which still contains a certain amount of water). The liquid portion of the mixture and the solids that are drained out of the scale with a certain delay are mixed to a slurry in crutchers or with some other type of forced mixer; the slurry is then transferred to a stirred storage vessel. Further processing can then occur in a continuous fashion. In addition to the batch slurry method described, continuous slurry preparation processes also exist.

Conversion of the slurry into powder requires the use of pressures up to 8.0 MPa (80 bar). The most practical means of slurry transport uses a slowly moving three-plunger pump, usually preceded by a low- or medium-pressure booster pump of either the rotary or positive-displacement variety. Between the storage vessel and the high-pressure pump, it is advantageous to provide a magnetic separator, sieves, and/or wet-grinding mills for removing metallic objects and large particles or agglomerates that might otherwise clog the spray nozzles. Bulk density can be increased by evacuating the slurry, whereas deliberate introduction of air can be used to reduce the density. Any sudden changes in pressure in the high-pressure portions of the system are compensated in an air vessel. Slurry is distributed in the tower head by a ring tube connecting a series of nozzle guns, each equipped with swirl nozzles. Atomization occurs as the slurry emerges from the nozzles. The number of nozzles and their type must be such that individual spray cones do not overlap, and the sides of the tower should remain clean. Dry tower powder emerges from the tower cone at a temperature of 90–100 °C, after which it is cooled to prevent clumping. An airlift is convenient for this purpose, particularly if vertical transport is necessary.

Air Flow. Heat and material transport in the hot air drying phase of the spray-drying tower process can best be described by the following model. Instantaneous evaporation of water from a spherical droplet into the surrounding heated air leads initially to solidification of the drop's outer shell, causing formation of a more or less plastic skin. As each spherical particle falls through the tower, additional moisture evaporates from its interior. This results in formation of a hollow body, described earlier as a bead, often with a surface fractured by the "craters" that are formed as water vapor escapes.

Some atomizing towers are constructed so that both product particles and air move in the same direction, in which case the ejected slurry emerges into the hottest region of the tower. This cocurrent process leads to rapid evaporation and extensive particle blowing, and the resulting beads are relatively light (Fig. 62 B). Residence time in the tower is brief and the product is removed from the "cold" region; hence, the process is regarded as rather gentle. Thus, it is particularly suited to the manufacture of detergents with a high surfactant content, such as those for easy-care fabrics.

Much more common for detergent drying, however, is a countercurrent process, in which the air and the product move in opposite directions. In this case, drying begins in a region of high humidity and lower temperature, proceeds at a slower rate than in the cocurrent process, and results in beads with thicker walls (cf. Fig. 62 C). The corresponding lower velocity of fall means a higher number of particles within the tower at any given time, a circumstance that encourages agglomeration. Thus, products from a countercurrent system are heavier and coarser than those from a system with a cocurrent arrangement.

The countercurrent operation illustrated in Figure 63 provides for dry air to be drawn in by a fan. The air is heated directly by the exhaust gas from an oil- or gas-fired burner and is led into the tower at 300 °C through a ring channel. Process economics and product quality depend both on maintaining adequate temperature and flow distribution throughout the tower cross section and on achieving the highest possible values for temperature and air flow rate consistent with a minimum of dust carryover.

Proper design of both the air inlet zone above the cone and the air outlet at the tower dome are crucial. Air entering the tower is deliberately "swirled"; i.e., it is accelerated tangentially as well as vertically, causing it to rise spirally. This helps ensure homogeneous thermal and material transfer between air and product and discourages the formation of convection currents. Gaseous water, combustion gases, and drying air are withdrawn from the tower by means of a ventilating fan. The capacities of the two fans are adjusted so that a slight negative pressure exists in the tower, thereby preventing escape of dust (e.g., in the vicinity of the nozzles).

The fines that are drawn out of the tower with the exhaust gas are collected in cyclones or, more effectively, in bag filters. Modern plants employ jet filters, in which dust is collected on the exterior of needled felt bags mounted on support baskets. These bags are cleaned by short blasts of compressed air from a Venturi nozzle in the head of the bag. Collected dust can be combined with the principal product as it exits from the tower, or if quality considerations dictate, it can be returned to the tower for re-agglomeration. Another more expensive alternative is to invoke wet-processing, where the dust is recycled to the slurry preparation stage.

Heat exchange systems are also important in detergent spray-drying plants [173]. For example, if heat exchangers are installed in the air inlet and outlet paths and are joined by a water circulating system, the resulting energy savings can be 10–20 %.

Details of layout and data for operation for any particular large manufacturing facility, of course, depend highly on both the mode of operation employed and the resulting product spectrum. Nonetheless, the following parameters can be regarded as typical for a countercurrent spray-drying tower:

construction material: carbon steel or stainless steel
diameter: > 6 m
cylinder height: ≤ 30 m
drying air requirement: 12–20 kg/kg of water
output capacity: ≤ 35 t/h tower powder
specific heat requirement (with heat recovery): 2900–4200 kJ/kg of water
tower inlet and outlet temperatures: 250–350 °C and 80–120 °C, respectively

Process Control. The quality of the powder produced in a spray-drying tower can be altered by adjustment of a number of parameters.

Variables with respect to the slurry are
quantity
pressure
temperature
air content

Variables with respect to the drying air are
quantity
inlet temperature
outlet temperature
axial velocity
tangential velocity

Product variables whose values are related to desired properties are
moisture
bulk density
particle size distribution
flowability

If process management is to be efficient, all of these variables must be measured and controlled accurately and, preferably, continuously. The most difficult part is determining the optimal parameters for processing the system, because the variables are mutually interdependent in a complex way; each can strongly influence others [174]. Modern process computers greatly simplify the task by providing statistically analyzed data [175]. If proper software is developed, computerization of the entire operation of a detergent spray-drying tower is possible. The advantages go far beyond increased automation, an end that might be achieved in other ways as well. More important is the information output, which can be used for further process development and corresponding gains in operational efficiency.

6.2.3 Finishing Process

Before tower powder can be marketed as a finished product, it must be supplemented with various ingredients. These include thermally sensitive materials such as sodium perborate, bleach activators, enzymes, and fragrances, and occasionally special surfactants and colored granules. Finishing of the tower powder can be executed immediately after the spray-drying process. The flow diagram shown in Figure 63 suggests incorporation of an intermediate storage step for tower powder. The crude material is first transported vertically and/or horizontally, using standard technology (belt conveyors, elevators, pneumatic pressure and suction transfer devices) [176], [177], after which the material is stored in appropriate supply bunkers. Subsequent removal is facilitated by designing the bunkers so that they are suited to the flow characteristics of the powder [178]–[180]. Installation of supplemental shakers and fluidizing nozzles can also be useful. The tower powder and residual components are then led over conveyor scales with alternating failure control, from which the components proceed along a collecting belt to an efficient mixer. This must be so designed that its action is efficient but relatively gentle to minimize abrasion [181]. Fragrances can be added by means of proportioning pumps located ahead of the mixer, although this step can also be a part of the mixing process. The finished product is then ready for sieving and packaging.

6.2.4 Spray-Mixing Process [182], [183]

As already described, powder prepared by a high-temperature spray-drying process must be submitted to a second stage of formulation to add heat-sensitive components. Therefore, efforts have been made to simplify the procedure; in particular, major advantages are found in replacing the thermal drying step partly or wholly with a procedure requiring less capital investment and less energy expenditure [184]. All such alternatives are here grouped under the general term "spray-mixing." In essence, they represent further developments on the crystallization technique for the preparation of dustless, granulated products.

Figure 64 is a block diagram of a generalized spray-mixing system; details vary from case to case.

Measured charges of solids and liquids are first premixed and are then led to the spray mixing apparatus. The amount of aqueous material that is permitted is limited by the absorption capacity of hydrate-forming components present, a value that is dictated by stoichiometric considerations. Thus, raw materials with a low water content from a neutralization step are sometimes used (see Section 6.3).

The necessary steps in the process include mixing of the solids, spraying these with the liquids, and agglomerating the whole into a dustless product. Appropriate mixing units can be designed to function on the basis of several principles:

gravitational settling of the solids in the course of rotational motion
movement of the powder by means of fixed inserts
creation of turbulence with an air stream

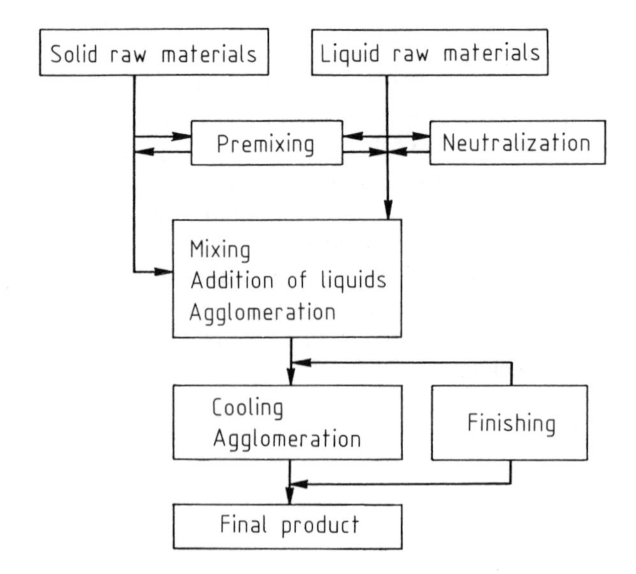

Figure 64. Schematic block diagram for spray-mixing manufacture of detergents

The liquids are sprayed onto a suspended powder screen so that even distribution and controlled agglomeration are encouraged. Crystallization is accelerated by removing heat that is released during the hydration process and either dissipating it to the mixing unit (e.g., in the air stream) or conducting it away by means of a separate cooling facility. Thermally unstable components such as perborate can be handled in one of three ways, depending on the extent of the heat evolution: (1) they can be added directly to the mix, (2) they can be introduced under milder conditions in a second processing zone, or (3) they may be reserved for later addition. Products prepared in this manner are generally quite homogeneous and low in dust content. Nevertheless, two problems inherent in any crystallization approach remain: (1) lack of flexibility with respect to raw material choice and (2) product bulk density. These matters are addressed separately in the form of specialized equipment options, available either as single components or as complete continuous process systems. For example, raw material flexibility can be enhanced somewhat by providing additional drying capacity. Other solutions involve a second step for agglomerating spray-dried partial formulations with residual raw materials. The suppliers of raw materials also provide help in the form of special products designed for spray-drying applications, including various premixed materials (e.g., in bead form so that they have a low bulk density).

The optimal approach for manufacture of a given product can only be determined after evaluation of all the analytical parameters outlined at the beginning of Section 6.1. [185]. An extensive review of various types of available equipment is contained in [182].

6.3 Raw Materials

The properties and functions of detergent components have already been described in Chapter 3. Some of these will be considered in this section from the standpoint of specific demands and potential roles with respect to the process of detergent manufacture.

6.3.1 Surfactants

In both the spray-drying and the high-pressure nozzle atomization processes, the moisture content of the surfactants must often be kept to a minimum. This is particularly true if predrying of other components is to be avoided. Low moisture content also permits complete utilization of water that is required for cooling and rinsing without sacrificing maximum slurry concentration.

Anionic surfactants are sometimes supplied as aqueous solutions. Materials such as sodium alkylbenzenesulfonate and soaps can be prepared by direct neutralization with 50% sodium hydroxide solution of concentrated forms of the corresponding sulfonic or fatty acids, the neutralization occurring within a slurry. If the spray mixing process is used for detergent manufacture, the neutralization can be carried out advantageously by deposition of the components on an alkaline carrier, or the substances can be sprayed through neutralizing mixer nozzles [184].

Nonionic alkyl and alkenyl polyglycol ethers exist in anhydrous form; thus, these materials are preferred for use in a spray-mix detergent formulation. Proper choice of carbon chain length and degree of ethoxylation must ensure compatibility of melting and solidification temperatures with the processing conditions to which the materials are exposed. With high-pressure nozzle atomization, many components of nonionic and anionic surfactants that are volatile in steam have a tendency to form organic aerosols in the exhaust gases. This can be controlled by proper choice of raw materials, adjustment of the atomization conditions, or treatment of the exhaust gases. Alternatively, these components can be introduced in a subsequent spraying and mixing operation involving the tower powder, for example, or some other suitable carrier material such as perborate [186].

The viscosity of a slurry is greatly influenced by the amount and type of surfactants present [187], [188]; gel formation can play a significant role. In the case of a complex mixture, optimization normally requires considerable experimentation and may involve varying the operating conditions, changing the material mixture, or adding hydrotropic substances.

6.3.2 Builders

Sodium Triphosphate ($Na_5P_3O_{10}$). The most important characteristic of this builder for finishing operations is its reaction with water. The reaction can be either a favorable one, hydration to give the hexahydrate, or an undesirable one, hydrolysis to di- and monophosphate [189]. Hydrolysis is most often encountered in drying towers operated at a high temperature, and it can be suppressed by lowering the temperature of either the slurry or the inlet air into the tower. The extent of hydrolysis is normally reported in terms of the degree of triphosphate retention.

The solids content of a slurry must be kept as high as possible if the process is to be economical; this optimization has certain consequences with respect to the proper choice of phosphate quality. The viscosity of a slurry increases continuously until all of the phosphate has been hydrated. Hydration rate can be influenced by a number of factors. For example, hydration is accelerated by increased content of phosphate in its high-temperature modification (phase I) as well as by increase in surface area, high purity, and the presence of hexahydrate seed crystals. Careful adjustment of these parameters enables manufacturers to offer phosphate of various types, and detergent makers can then choose a material whose hydration rate and viscosity match the demands imposed by their particular production facilities [190].

For the spray-mixing process, the phosphate industry offers special grades of phosphate that have a bead structure and, thus, a low bulk density.

Sodium Aluminum Silicate (Zeolite 4 A) [191]–[195]. Awareness of increasingly serious eutrophication problems in surface waters stimulated a vigorous search for phosphate substitutes, and thereby the effectiveness of certain water-insoluble sodium aluminum silicates as builders was discovered. The application and washing activity properties of such materials have already been discussed in Chapter 3. Commercial use of insoluble silicates in detergents first became possible after research efforts were successful in the development of the manufacturing process for zeolite 4 A with its specific particle size distribution characteristics. Products of this type are totally removed from fabrics by the rinse cycle which follows a normal washing. Large-scale manufacture of zeolite 4 A for detergent use has existed only since 1978; thus, the process is discussed here in some detail [195]:

Manufacture. Zeolite 4 A can best be described by the overall formula $Na_{12}(AlO_2)_{12}(SiO_2)_{12} \cdot 27\,H_2O$, which is also the formula of its unit cell. One way in which it can be prepared is by first converting kaolin into metakaolin and then transforming the latter into zeolite by treatment with alkali:

$$Al_2O_3 \cdot SiO_2 \cdot 2\,H_2O \xrightarrow[-2\,H_2O]{Heat} Al_2O_3 \cdot SiO_2 \xrightarrow[NaOH]{Heat} Na_2O \cdot Al_2O_3 \cdot 2\,SiO_2\,(aq) \text{ Zeolite 4 A (empirical formula)}$$

However, a more important process currently is the hydrothermal treatment of sodium silicate with sodium aluminate in the presence of alkali:

$$Na_2SiO_3\,(aq) + NaAlO_2\,(aq) \xrightarrow[NaOH]{Heat} Na_2O \cdot Al_2O_3 \cdot 2\,SiO_2\,(aq)$$

A suitable facility for zeolite 4 A precipitation and crystallization is shown schematically in Figure 65. Raw materials are introduced within defined concentration ranges in a gravity tank, where with rising viscosity they form a gel having a high water content. Precipitation is carried out in relatively dilute solution with intensive stirring. Batch operation can also be replaced by a continuous process that uses multistage stirring columns.

Subsequent transformation of the crude product into the highly ordered crystalline mass of zeolite 4 A occurs in a crystallization vessel when the temperature is increased. After crystallization is complete, the mother liquor is removed by passing the mixture over a suitable separator (e.g., a vacuum belt filter) to give a filter cake with a product content of ca. 40–50 %. This filter cake is then washed. The mother liquor and wash water can be recycled after they have been concentrated in an evaporation step as far as required. Evaporator condensate can then be used as wash water.

Crude product in the form of the filter cake can be further processed for detergent purposes by spray-drying; alternatively, additives can be introduced to stabilize the product as a suspension [196].

The dried material is a fine powder. It has a moisture content of ca. 20 % and a bulk density of ca. 350 g/L. It lends itself well to standard modes of transport and storage in silos.

When detergents are prepared by spray-drying [197], zeolite 4 A is commonly introduced in the form of a suspension (density 1.1–1.5 g/cm^3). This may have both commercial and economic advantages. The suspension is stabilized to prevent zeolite

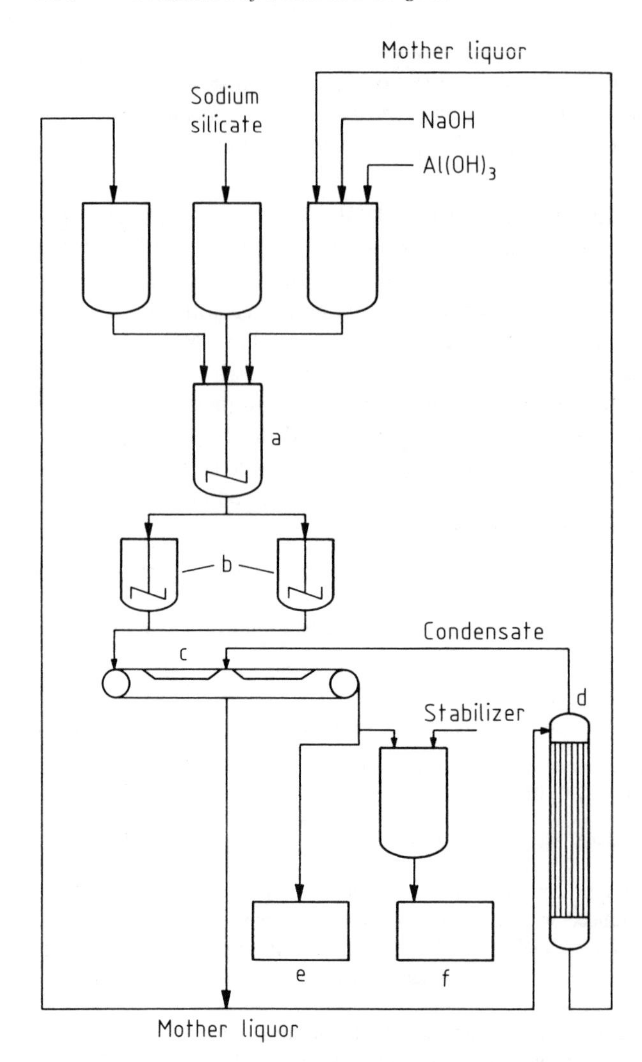

Figure 65. Flow diagram for the preparation of zeolite 4A a) Precipitation; b) Crystallization; c) Filtration; d) Evaporation; e) Zeolite 4A spray-drying; f) Zeolite 4A stabilized suspension

sedimentation during transport and storage and to ensure proper viscosity. Suitable stabilizers include a small amount of compounds that are themselves standard detergent components (e.g., specific non-ionic surfactants). The viscosity of the suspension is influenced by concentration, temperature, particle size distribution, stabilizer content, and the magnitude of shear forces to which it is subjected.

Replacement of sodium triphosphate by zeolite 4A need not result in a change in the solids content of a detergent slurry. The water content of the zeolite suspension can be compensated by supplying other raw materials in more concentrated form. The described increase in viscosity of a triphosphate-containing slurry is suppressed. Tower drying characteristics of a zeolite-containing slurry differ from those observed with a pure phosphate preparation, which requires that drying conditions also be adjusted accordingly.

The importance of phosphate retention values has already been mentioned in the context of sodium triphosphate. These are also a matter of concern with detergents employing zeolite 4A–sodium triphosphate builder mixtures. The fact that zeolite 4A confers excellent flow characteristics makes possible preparation of zeolite-containing tower powder with a relatively high moisture content. This reduces the thermal stress to which the accompanying phosphate is subjected and helps compensate for the increased decomposition that is normally encountered when phosphate content is lowered.

In principle, zeolite 4A can also be utilized in a spray mix operation. Nonetheless, this practice raises problems, particularly if high zeolite concentrations are planned and if the facility lacks provisions for postdrying. Thus, one must bear in mind that ca. 45% of the total raw material mass is due to water, and all of this must be absorbed by other hydrate-forming components of the mix. For this reason it is advisable to employ either spray-dried raw materials or special zeolite 4A compounds.

6.3.3 Other Detergent Raw Materials

Sodium Silicate. Soluble sodium silicates with a $Na_2O:SiO_2$ mole ratio between 1:2 and 1:3.4 enhance the stability and flowability of a washing powder. These materials are suitable for effective agglomeration, which is intended to occur in a spray-mixing process.

Sodium Perborate. Heavy-duty detergents marketed in Western Europe usually contain sodium perborate in the form of its tetrahydrate. For reasons of stability, the temperature should not be permitted to exceed 60 °C. Perborate can serve very effectively as the carrier for nonionic surfactants if, for reasons described above, these prove inconvenient to introduce at the slurry stage.

Enzymes. The majority of the enzymes that are used in detergents display proteolytic activity. When enzymes were first introduced into detergents in the 1960s, their physiological properties caused problems. Most of the problems were solved by the institution of improved production safety procedures, as well as by the development of virtually dust-free surface-sealed enzyme preparations in the forms of prills or pellets, which can be handled with complete safety [198].

7 Analysis

The fact that detergents can consist of a large number of individual components whose structures vary greatly is apparent from Chapter 3. The dramatic industrial advances of the past 30 years coupled with ecological and economic pressures have resulted in nearly constant modification of detergent formulas. This modification has required steady development in analytical approaches, with routine procedures regularly becoming outdated.

The multitude of products and possible raw materials, as well as the complexity of their constitution, makes a summary of the techniques of detergent analysis difficult, especially since the situation is complicated even further by the wealth of available analytical techniques.

The most important factor in choosing an analytical method is the nature of the question posed. In addition, the value of an analysis is highly dependent on the use of proper sampling techniques. Often several methods of analysis are employed in the search for the single answer.

The following procedure can be regarded as a typical example of how a general analysis of a powdered detergent might be conducted:

First, the product is separated by extraction into ethanol-soluble and ethanol-insoluble fractions. The ethanol extract can usually be regarded as containing all of the surfactants. The various surfactants are then determined: anionic surfactants by two-phase titration, nonionic surfactants by passing the ethanol extract through both anionic and cationic ion-exchange resins, and the soaps either by titration of the alkalinity of the ethanol extract or by a second two-phase titration in alkaline medium. After suitable sample preparation, such major inorganic components as carbonates, silicates, phosphates, borates, and sulfates are determined either instrumentally or by classical wet methods. All of the more specialized components (e.g., soil antiredeposition agents, fluorescent whitening agents, enzymes, chelating agents, bleach activators, etc.) are present in relatively small quantities, and their analysis requires considerable knowledge and experience. Thus, they would fall outside the bounds of a general investigation.

Following is a summary of the literature on detergent analysis. This is intended only as a useful overview. The information has been divided into five separate categories:

1. Detergent ingredients
2. Purposes of detergent analysis
3. Importance of sample preparation
4. Analytical methods
5. Sources of information

7.1 Detergent Ingredients: [199]–[209]

Inorganic components: [210]–[217]
Organic components: [218]–[221, Appendix [1–21], [222], [223]

7.2 Purposes of Detergent Analysis: [224]

Raw material analysis (triphosphate and alkylbenzenesulfonate): [225]–[227]
Production control (EO products): [228]
Product control (detergent analysis): [229], [230]
Market analysis (detergents): [221], [222, pp. 991–1045], [231]–[236], [237, pp. 25–43 (1977)]
Trace analysis (vinyl chloride in PVC): [238]–[240]

7.3 Sample Preparation: [221, pp. 21–26]

Liquids, solutions, suspensions, emulsions, and pastes: [226, p. 6], [241, H–I 4 (65)]
Bar detergents [241, G–I 1–3 (50)], [242, Da 1–45]
Powders: [241, H–I 5], [242, Dc 1–59, Dd 1–59], [243, ISO 607 (1980)], [244], [245, DIN 53911]

7.4 Analytical Methods

7.4.1 Qualitative Analysis: [221, pp. 27–49], [233, pp. 166–178]

General Analysis:
Solubility, pH: [241, H–III 1 (65)]
Elemental analysis: [241, H–III 5 (65)]

Analysis for Inorganic Ingredients: [246], [247]
Active oxygen: [246, p. 171], [247, pp. 492–493]
Active chlorine: [246, p. 159], [248, Part 2, vol. VII, p. 35]
Ammonia: [246, p. 241], [249, E 5]
Boron: [246, p. 230], [247, p. 776],
Water-insoluble inorganic builders: [204], [250], [251]

Analysis for Organic Ingredients: [252]–[254]
Alkanolamines after hydrolysis: [221, p. 32], [223], [255]
Sequestrants (NTA, EDTA): [256], [257]
Urea: [258], [259, p. 235], [260]
Fluorescent whitening agents: [221, p. 414], [261]–[263]
Anionic surfactants: [264]–[267]
Alkylene oxide adducts (EO, PO, aromatics): [221, p. 46], [252, p. 138], [266], [268], [269]
Hydrotropic substances: [258]
Enzymes: [270]
Carboxymethyl cellulose: [271], [272]

7.4.2 Sample Preparation

Extraction, Perforation, and Sublation
Ethanol solubles: [221, p. 148], [241 H–III 4 (65)]
Soap fatty acids: [241, G–III 6b (57) and G–III 8 (61)], [273, p. 1352]
Monosulfonate in alkanesulfonate: [226, p. 39], [241, H–IV 2b (76)], [243, ISO 3206 (1975)]
Acetone solubles: [221, pp. 253 and 362]
Nonionic surfactants in wastewater: [274]
Nonionic surfactants in sludge: [275]
Polyglycol in ethylene oxide adducts: [276]

Ion Exchange: [221, pp. 50–58], [277]
Nonionic surfactants, anionic surfactants, cationic surfactants, soap fatty acids, and urea: [278]–[285]
Antimicrobicides: [286]
Phosphate types: [221, p. 467]

Carbon Methods
Sulfate: [241, H–III 8b (76)]
Phosphate: [287]

Decomposition Methods: [288]
Residue on ignition, with hydrofluoric acid digestion: [289, p. 8]
Sulfate ash: [221, p. 335], [241, H–III 11], [289 a]
Borate: [221, pp. 474–476]
Borate decomposition for X-ray fluorescence analysis: [227], [290], [291]
Potassium disulfate melt for Al and Fe: [289, p. 12]
Soda–potash decomposition for water glass: [289, p. 9]
Wet Decomposition
Sulfuric acid–nitric acid, perchloric acid, nitric acid–perchloric acid: [292, pp. 188–209]

Chemical Transformations: [293]
Sodium hydrogen sulfite–acid hydrolysis: [221, p. 152], [241, H–IV 7 a], [293 a], [273, pp. 1474 and 1353], [294], [295]
Cleavage of ether sulfates: [221, pp. 279 and 323], [296]–[301, D 855–6]
Esterification of fatty acids: [221, p. 572], [302]–[305]
Desulfonation (alkylbenzenesulfonates, alkanesulfonates): [221, p. 197], [297], [298], [306]–[311]
Acetylation (fatty alcohols): [312, p. 543]

Alkaline Hydrolysis: [221, p. 150]

7.4.3 Quantitative Analysis

Gravimetric Analysis
Inorganic Ingredients
Dry residue: [221, p. 434], [313]
Carbon dioxide: [248, Part 3, vol. IV aα pp. 225–229]
Silicon dioxide: [301, D 501–558], [314], [221, p. 447], [248, Part 3, vol. IV aα, p. 456]
Phosphorus pentoxide: [221, p. 452], [243, ISO/TC 91 444F], [248, Part 3, vol. V aβ, p. 118], [315]
Sulfate: [241, H–III 8 b (76)], [316]
Magnesium: [221, p. 492], [248, Part 3, vol. II a, pp. 141–149]
Organic Ingredients: [273]
Ethanol solubles: [221, p. 148], [241, H–III 4 (65)]
Loss on drying: [221, p. 548], [313]
Unsulfated fraction (US): [221, p. 555], [243, ISO–893 (78) and ISO–894 (77)]
Nonsaponifiable fraction: [221, p. 555], [241, C–III 1 a and 1 b (77)]
Soap fatty acids: [221, p. 550], [241, G–III 6 b (57) and G–III 8 (61)], [273, p. 1352]
Nonionic surfactants: [221, p. 167], [241, H–IV 9 b]

Volumetric Analysis: [221, p. 215], [317], [318]

Acidimetry

Alkalinity: [221, pp. 334 and 472]
Sodium oxide content in soaps: [221, p. 554], [241, H–III 7a (65)], [319]
Borate: [221, p. 474], [248, Part 3, vol. III, pp. 11–60]

Potentiometry: [320]–[322]
Phosphate: [221, pp. 455 and 480]
Borate: [221, p. 480], [323]
Chloride: [221, p. 446], [248, Part 3, vol. VII aβ, pp. 107–109]
Fluoride: [324], [325]
Nonionic surfactants in wastewater: [274]

Complexometry: [326], [327]
Ethylenediaminetetraacetate: [221, pp. 403 and 487], [243, ISO TC 91/490]
Nitrilotriacetate: [221, p. 403], [243, ISO TC 91/513 E]
Citrate: [328]
Calcium: [249, E 3], [329, p. 25]
Magnesium: [326, pp. 136–137], [329, p. 28], [249, E 4]
Nonionic surfactants: [330]

Conductometry: [331]
Sulfate: [332]

Two-Phase Systems: [221, pp. 215 and 218], [231, p. 427], [226, p. 16], [333, p. 221]
Anionic surfactants: [221, p. 234], [241, H–III 10], [333 a], [295], [334]
Cationic surfactants: [243, ISO 2871 (73)], [335]
Turbidity titration: [336]

Miscellaneous Titrimetric Analyses

Sulfate: [221, p. 442], [226, H–III 8a (76)], [241, H–III 8a (76), p. 24; H–III 8d (76)], [337], [338]
Chloride: [221, p. 445], [248, Part 3, vol. VII aβ, pp. 98, 99]
Water by Karl Fischer method: [221, p. 435], [241, H–III 3a (65)], [243, ISO 4317 (77)]
Active oxygen: [221, p. 483], [243, R 607], [248, Part 3, vol. VI aα, pp. 193–207]
Active chlorine: [221, p. 487], [248, Part 3, vol. VII aβ II, pp. 74–76]

Spectrophotometric Determinations: [292]
Perborate: [339, A 124]
Phosphate: [221, p. 457], [340], [341]
Enzymes: [342], [343]
Enzymatic analysis: [344]
Trace elements: [292]
Anionic surfactants in wastewater: [316], [345]–[347]
Cationic surfactants in wastewater: [348]

Flame Photometry: [349]
Sodium and potassium: [349]

Atomic Absorption Spectrometry: [350]–[352]
Traces of Cu, Fe, Zn, and Mn: [353], [354]

X-Ray Fluorescence: [227], [355]–[357]

X-Ray Diffraction: [358]
Phosphate phases I and II: [359]
Zeolite types: [360]

Radiometric Methods: [361]

7.4.4 Separation Methods: [362], [363, p. 21]

Paper Chromatography: [364]
Phosphate types: [221, p. 460], [225], [365]
Alkanolamines: [366]
Alkanolamides: [221, p. 391]
Urea: [362, pp. 360 and 402], [364, p. 210], [367]

Thin-Layer Chromatography: [221, p. 45], [254], [368], [369]
NTA, EDTA: [370]
Akanolamines: [223]
Ether sulfates: [371]
Surfactant mixtures: [266], [267], [333, p. 47], [372]
Free fatty alcohols in ethylene oxide adducts: [373], [374]
Bactericides: [286], [375]
Amphoteric surfactants: [221, p. 371], [376]
Anionic surfactants: [377]–[379]
Nonionic surfactants in wastewater: [380]

Column Chromatography: [221, pp. 51–63], [363], [381], [382]
Glycerides [273, Part I, p. 832], [383]
Fatty acid polyglycol esters: [384]
Petroleum sulfonates: [221, p. 61], [385]
Nonionic surfactants: [221, p. 167], [386]

Gas Chromatography: [221, p. 80], [387]–[390]
Fatty acid carbon-chain distribution: [305], [391]–[393]
Alkylbenzenesulfonate chain distribution: [394]

Fatty alcohol carbon-chain distribution: [395]
Solvents from cleansers: [396], [397]
Propellants in sprays: [398]–[400]
Dioxane in ethylene oxide adducts: [401]
Antimicrobial agents: [375]
Alkoxy content of cellulose ethers: [272]

Liquid Chromatography: [389], [402]–[404]
Poly(ethylene glycol) in nonionic surfactants: [405]–[407]
Unreacted alkylbenzenes and fatty alcohols in their sulfonation or esterification products: [408]
Separation of fluorescent whitening agents: [409]
Bacteriostatic substances in soaps: [410]
Nonionic surfactants: [389], [411]
Anionic surfactants: [412], [413]
Separation of sulfonic acids: [414], [415]
Separation of lower alcohols: [416]

7.4.5 Structure Determination

UV Spectrometry: [221, p. 74], [417]–[420]
Alkylbenzenesulfonates: [242, Dd 3–60], [421]–[423]
Toluenesulfonate, xylenesulfonate: [221, p. 205], [424]
Structural information of surfactants: [385], [425]
Toluenesulfonamides: [426]

IR Spectrometry: [221, p. 69], [233], [427]–[433]
Petroleum sulfonates: [385]
Degree of branching in alkylbenzenesulfonates: [434], [435]
Surfactant sulfonates: [436], [437]
Soil antiredeposition agents, cellulose derivatives: [438]

NMR Spectrometry: [221, p. 86], [439], [440]
Ethylene oxide–propylene oxide adducts: [221, p. 301], [389], [441], [442]

Mass Spectrometry: [221, p. 76], [443]–[445, pp. 1–46], [446]
Fragrances (GC/MS): [447], [448]
Bactericides (TLC/MS): [221, p. 426], [449]

7.4.6 Determination of Characteristic Values

Elemental Analysis: [450]
Nitrogen: [221, pp. 124 and 444], [450, p. 178], [451]
Halogen: [221, p. 128], [452], [453]
Sulfur: [221, pp. 126 and 439], [450, pp. 252 and 307], [453]
Phosphorus: [221, p.129], [292, pp. 899 and 902], [450, p. 360]

Fat Analysis Data: [273], [454]
Acid number: [221, p. 182], [241, C–V 2 (77)]
Saponification number: [221, p. 183], [241, C–V 3 (57)]
Iodine number: [221, p. 183], [241, C–V 11 a–d (53)]
Hydrogen iodide number: [241, C–V 12 (53)]
Hydroxyl number: [221, pp. 186 and 303], [241, C–V 17a + b (53)]

7.4.7 Analysis Automation: [455]

Phosphorus pentoxide: [456]–[458]
Proteolytic enzymes: [459]

7.5 Sources of Information: [237, p. 403 (1981)]

Analytical Commissions: [460], [461]
Federal Republic of Germany
GAT (Gemeinschaftsausschuß für die Analytik von Tensiden) [461 a]; member organizations of GAT:

a) DGF (Deutsche Gesellschaft für Fettwissenschaften) Soester Straße 13, 4400 Münster, Federal Republic of Germany
b) DAGSt (Deutscher Ausschuß für Grenzflächenaktive Stoffe im VCI); this organization was disbanded in 1978. Its duties have been assumed by the TEGEWA (Verband der Textilhilfsmittel-, Lederhilfsmittel-, Gerbstoff- und Waschrohstoff-Industrie e.V., Karlstraße 21, 6000 Frankfurt/Main 1) and the IKW (Industrieverband Körperpflege und Waschmittel e.V., Karlstraße 21, Frankfurt/Main 1).
c) NMP (Normenausschuß Materialprüfung im DIN) Burggrafenstraße 4–10, Postfach 11 04, 1000 Berlin 30

International Organizations
CIA (Commission Internationale d'Analyses of the CID) was disbanded in 1978. Its work has been carried on by the CESIO/AIS (Comité Européen d'Agents de Surface et Intermédiaires Organiques/Association Internationale de la Savonnerie et de la Détergence) Working Group on Analysis: [462], [463]
ISO (International Organization for Standardization)
IUPAC (International Union of Pure and Applied Chemistry) Commission on Oils, Fats, and Derivatives in the Applied Chemistry Division

Collections of Methods, and Sources from Which They May Be Obtained:

DGF-Einheitsmethoden.
Single pages: Deutsche Gesellschaft für Fettwissenschaften e.V., Soester Straße 13, 4400 Münster, FRG.
Collections: Wissenschaftliche Verlagsgesellschaft mbH, Stuttgart, FRG.

DIN Standards.
Deutscher Normenausschuß e.V. (DNA), Beuth Verlag GmbH, Burggrafenstraße 4–10, 1000 Berlin 30, FRG.

IUPAC Standard Methods.
International Union of Pure and Applied Chemistry, Butterworth & Co., Kingsway WC 2 B 6 AB, London 88, United Kingdom.

ASTM Standards.
American Society for Testing and Materials, European distributor: Heyden & Son GmbH, Münsterstraße 22, 4440 Rheine, FRG.

AOCS Official and Tentative Methods.
American Oil Chemists Society, 508 South Sixth Street, Champaign, Illinois 61820, United States.

ISO International Standards.
Secrétariat ISO/TC 91 through Association Française de Normalisation, Tour Europe-Cedex 7, 92080 Paris La Défense, France.

8 Test Methods for Laundry Detergents

All detergent testing methods have the establishment of product quality as their goal. "Quality" is taken to include all those subjectively and objectively measurable properties of a product that play a significant role in its application.

Many possible approaches are available for measuring detergent quality, but these can be subdivided into three major groups:

laboratory methods
practical evaluation
consumer tests

8.1 Laboratory Methods

Practical evaluations require careful statistical analysis of increasing numbers of test series, most of which are time-consuming and associated with high labor and material costs. For this reason, development of preliminary test procedures has become necessary as well as suitable laboratory methods for obtaining useful and relevant information especially for development work. Such tests are designed to approximate service conditions to which the products will actually be subjected. The tests generally produce valuable clues of product quality, but their results should not be overrated, since the circumstances under which they are obtained are not identical to those that apply in the field. Normally these tests are conducted with laboratory equipment specially designed for small-scale tests (e.g., Launder-ometer [464], Linitest equipment [465], Terg-o-tometer [466], and foam testing equipment [467]–469]).

In place of naturally soiled laundry, use of artificially soiled fibers and test swatches is standard in laboratory tests. These swatches are prepared by using various fabrics and various soils. The materials to which the soil is applied are often commercially available [470], but some are prepared in the laboratory for specific purposes. Artificially soiled fibers or fabrics must be prepared so that they respond selectively to different detergent components and different sets of washing conditions if adequate differentiation is to be assured.

Criteria for testing detergents are subdivided as follows:

single wash cycle performance (soil removal and bleaching)
multiple wash cycle performance, e.g., after 25 or 50 washes (soil antiredeposition properties, degree of whiteness, buildup of undesirable deposits, fiber damage, stiffness, color change, fluorescent whitening)
special characteristics (powder characteristics such as density, free flowability, flushing by a washing machine, homogeneity, dusting properties, solubility, foaming, rinse behavior, and such storage characteristics as stability, hygroscopicity, color, odor, and tendency to clump)

The literature describes numerous methods for testing according to the above criteria, some of which are standardized. Standardization is a concern not only of national bodies (e.g., ANSI, the American National Standards Institute; JISC, the Japanese Industrial Standards Committee; DIN, Deutsches Institut für Normung [471]; AFNOR, Association Française de Normalisation; BSI, British Standards Institution), but also of international groups (e.g., ISO, International Organization for Standardization).

The above national organizations are all members of ISO and can, therefore, exercise influence on questions of international standardization [472]. Another particularly important organization concerned with international standardization of test methods is the CID (Comité International des Derivés Tensio-Actifs) with its subcommittee, the CIE (Commission Internationale d'Essai). This organization was disbanded in 1978, but in the meantime, its activities have been taken over and carried forward by the Working Group TMS (Test Methods for Surfactants) of the CESIO (Comité Européen d'Agents de Surface et Intermédiaires Organiques).

8.2 Practical Evaluation [473]–[477]

Practical wash evaluation is conducted with the exclusive use of commercial washing apparatus. In contrast to pure laboratory investigations, this work is carried out with laundry whose soil has been acquired naturally. The purpose of such experiments is to obtain results that can be regarded as realistic. Nonetheless, it is impossible to encompass the entire spectrum of conditions that might be encountered in practice. In many cases, so-called anonymous laundry is tested: laundry of household origin, but not otherwise identified. Far better, however, even though more difficult to arrange, are wash and use tests conducted by using specific new fabric items that have been distributed for everyday use in a sufficient number of representative households. Such samples are then reassembled at some central point and washed with the detergent being evaluated. The same item must always be washed with precisely the same detergent. The number of necessary wash cycles depends on the nature of the questions being investigated. Nonetheless, usually ca. 20–25 such use and wash cycles are necessary, occasionally even more. The detergent is generally used according to the manufacturer's directions, although effects of under- and overdosing must also be taken into account. Analytical precision is improved if instrumental methods are used to evaluate test results, but such measurements are often difficult both to interpret and to conduct. Thus, a better evaluation is based primarily on visual analysis, with physical measurements used only in a supporting role. Recommendations for comparative testing of detergent use characteristics have been the subject of a tentative DIN standard [478], as well as an ISO standard [479].

In the interest of both manufacturers and especially consumers, comparative use tests must be conducted under realistic conditions. This is particularly true of tests designed for consumer education, such as those carried out by consumer testing agencies; otherwise the results are of little value. Oversimplified tests incapable of making the sorts of distinctions that consumers consider relevant are a disservice to all concerned. Ultimate product judgements must be based not only on wash results, but on all of the relevant factors affecting use value.

8.3 Consumer Tests

To obtain a final degree of certainty with respect to a detergent's use characteristics and consumer acceptability, consumer tests are normally conducted. In such tests, consumers subject a product to their own sets of conditions; i.e., they use their own washing machines on their own laundry and soil, and it is assumed that they introduce the detergent in the manner to which they are accustomed. The precision of such tests is of course limited; therefore, it is necessary that a large number of consumers participate if statistically meaningful conclusions are to be drawn. Experience has shown that such tests are more valuable than any laboratory results, both for discovering product weaknesses and for identifying particular strengths. By their very nature, consumer tests subject a product to a great many different sets of circumstances, and they lead to the most meaningful results. Thus, consumer tests are very frequently conducted prior to the marketing of a new product.

9 Economic Aspects

The worldwide economic role of detergents represents a factor of considerable importance. Nonetheless, detergent consumption varies tremendously from country to country (Table 40) [480]. An examination of changing patterns of use for detergents and cleansers over the last 20 years reveals a remarkable growth in overall consumption, with the absolute quantity rising from 10.8×10^6 t in 1960 to 31.5×10^6 t in 1984.

Only a handful of companies are responsible for the production of most of the world's detergents and cleansing agents. The leader is Cincinnati-based Procter & Gamble, with Unilever of London second, while Colgate Palmolive in New York is third and Henkel in Düsseldorf, Federal Republic of Germany, is fourth. In the United States market, four firms (Procter & Gamble, Colgate, Lever, and Purex Industries) totally dominate, but internationally these four account for less than half the market. The remainder — ca. 53–55% — is split among many small local or regional producers. In Europe, for example, the Association Internationale de la Savonnerie et de la Détergence (AIS), a trade association made up of European national detergent industry groups, estimates that its members represent more than 700 individual firms.

Table 40. World consumption of soaps, detergents, and cleaning compounds [482]

Region	1960		1970		1980		1984	
	1000 t	Per capita, kg	1000 t	Per capita, kg	1000 t	Per capita, kg	1000 t	Per capita, kg
Western Europe	3047	9.7	4767	13.8	7050	18.9	7800	20.7
Eastern Europe	2034	6.5	2253	6.5	3268	8.6	3650	9.4
North America	2521	12.8	4574	20.2	7564	30.1	8200	31.3
Central America	247	4.1	617	6.9	2757 (Central and South America)	7.8	3300	8.3
South America	652	4.6	887	4.7				
Oceania	166	13.1	237	15.4	300	16.1	350	17.5
Africa	539	2.3	780	2.3	1494	3.2	1900	3.5
Asia	1594	1.0	2324	1.2	5241	2.1*	6250	2.3
World	10800	3.8	16439	4.6	27674	6.3**	31450	6.6

* Excluding China: 2.5 kg. ** Excluding China: 7.7 kg.

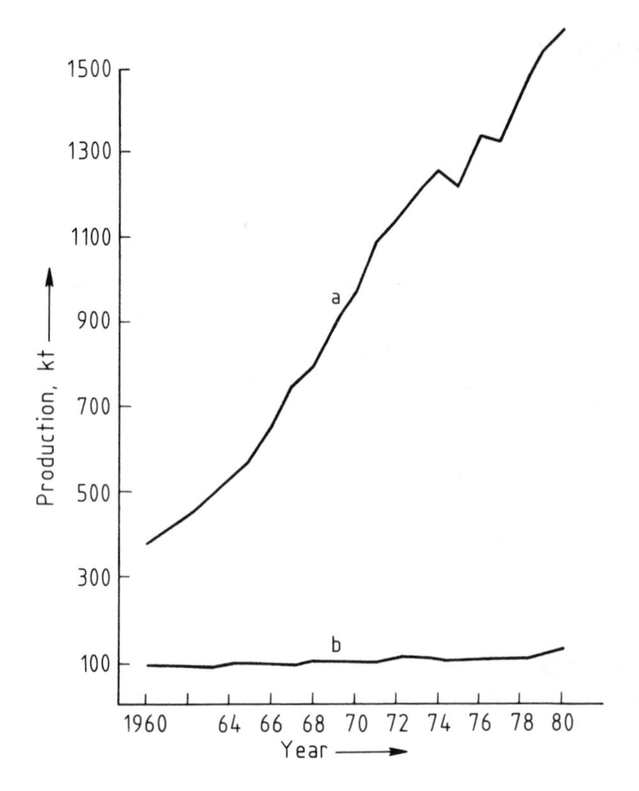

Figure 66. Production of soaps, detergents, and cleansing agents in the Federal Republic of Germany [481]
a) Synthetic detergents and cleansers; b) Soaps

The four largest firms do most of their business near their home markets. For example, ca. 70% of Procter & Gamble's sales occur in the United States. Similarly, Europe accounts for ca. 65% and 80% of the business of Unilever and Henkel, respectively. Only Colgate reports less than half of its sales at home: ca. 40% in the United States, with the balance split nearly equally between Europe and the remainder of the world [144].

The production and sale of soaps in some countries has developed at a very different pace from that of detergents and cleansers based on synthetic surfactants, as illustrated in Figure 66 for the Federal Republic of Germany.

9.1 Detergent Components

Production capacity and demand for the major raw materials of detergents are briefly reviewed in the following sections.

9.1.1 Surfactants

In 1982, 36% of the surfactants consumed in the United States, Western Europe, and Japan were used for washing and cleansing purposes (Table 41) [482]. The regional distribution of demand for surfactants is shown in Figure 67, while Figure 68 illustrates differences in the relative importance of various specific applications from one market to another. It is clear from these data that the two major sectors of the surfactant market are household and industry. In the three major market areas (United States, Western Europe, and Japan), the ratio in 1982 for anionic:nonionic:cationic surfactants was approximately 8:3:1 (Table 42).

Figure 69 provides an overview of the recent development of demand for surfactants in the United States, Western Europe, and Japan.

A breakdown for household consumption of various classes of surfactants in Western Europe, Japan, and the United States during 1982 is provided in Figure 29 (Section 3.1).

9.1.2 Builders

The most important detergent builder after World War II was sodium triphosphate. Its use has become a subject of intense public discussion in the last 15 years

Table 41. Consumption of surfactants vs. fields of applications in the United States, Japan, and Western Europe in 1982 [482]

	Application fields	Consumption, 1000 t	%	
Consumer products	detergents and cleaners	1900*	36	57
	cosmetics	1100*	21	
Industrial applications	foods	200	4	43
	plant protection and pest control	100	2	
	textiles and fibers	750	14	
	metal processing	130	2	
	paints, lacquers, and plastics	200	4	
	cellulose and paper	100	2	
	leather and furs	50	1	
	building construction	50	1	
	mining, flotation, and oil production	300	6	
	others	400	7	
Total		5280	100	

* Soaps included.

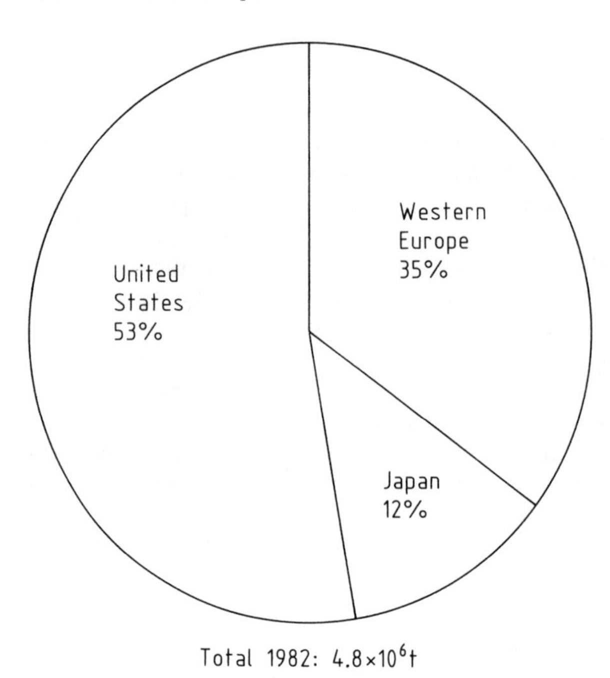

Figure 67. Surfactant demand in 1982 [42]

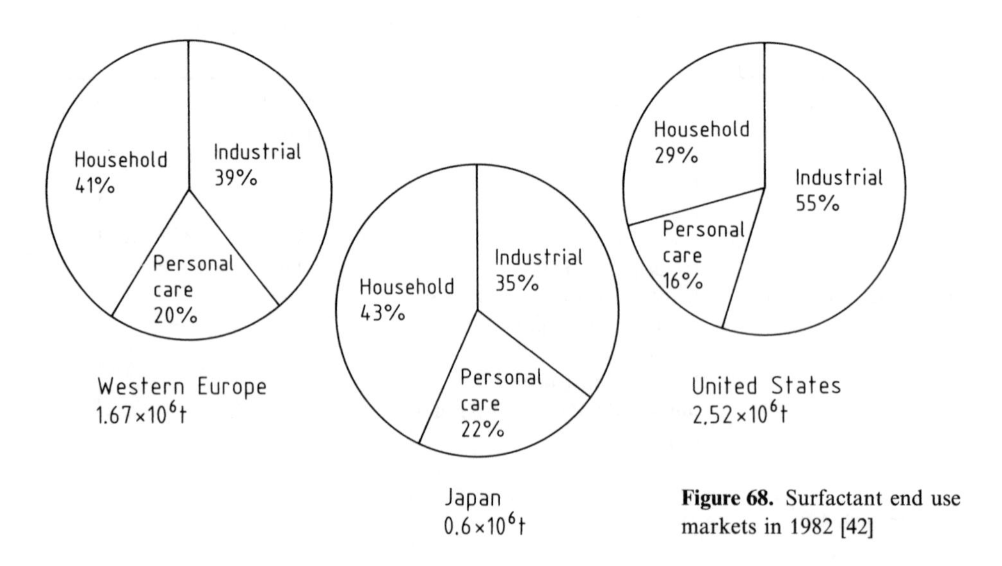

Figure 68. Surfactant end use markets in 1982 [42]

Table 42. Production of synthetic surfactants in the United States, Japan, and Western Europe in 1982 [483]

Surfactants		Quantity, t
Anionic surfactants (66.5%)	LAS	1 100 000
	ligninsulfonates	600 000
	fatty alcohol ether sulfates	360 000
	fatty alcohol sulfates	220 000
	petroleum sulfonates	140 000
Nonionic surfactants (25.0%)	fatty alcohol poly(ethylene glycol) ethers	500 000
	alkylphenol poly(ethylene glycol) ethers	310 000
	fatty acid amides and alkanolamides	100 000
Cationic surfactants (8.5%)	dialkyldimethylammonium chlorides and others	310 000
Total		3 640 000

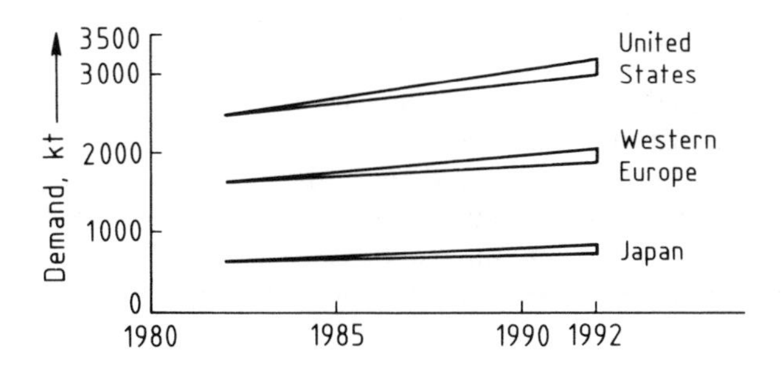

Figure 69. Surfactant demand growth 1982–1992 [42]

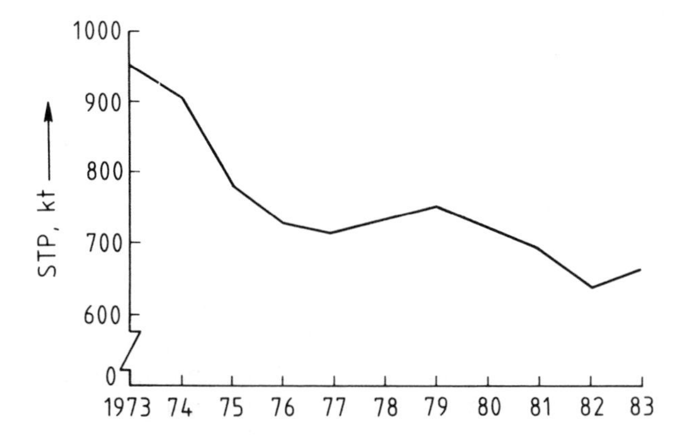

Figure 70. Production of sodium triphosphate (STP) in the United States [144]

Table 43. Consumption of sodium triphosphate in detergents, dishwashing agents, and cleansers during 1975 [484]

Application	Quantity, t
Household laundry detergents	240 000
Commercial laundry detergents	20 000
Automatic dishwashing agents	12 000
Household cleansers	4 000
Total	276 000

as a result of phosphate-induced eutrophication of surface waters. In countries where eutrophication has been particularly severe, agreements and legal restrictions which have led to a sharp decrease in the use of sodium triphosphate for washing and cleansing purpose have resulted. An example of the change in the pattern of use in the United States is shown in Figure 70. Sodium triphosphate use for washing in the Federal Republic of Germany fell from 260 000 t in 1975 (Table 43) [485] to only ca. 140 000 t in 1985. In Japan the reduction was even more drastic. Meanwhile, world-wide production capacity has risen dramatically for the increasingly important phosphate-substitute zeolite 4 A (Fig. 71). At the present time, for example, more than 90% of all detergents produced and sold in Japan are free of phosphate, containing zeolite instead. In those parts of the United States in which phosphate use is restricted or prohibited, nearly 50% of all powdered heavy-duty detergents contain zeolite. Virtually all of the powdered heavy-duty detergents made by one of the leading detergent manufacturers in Europe contain zeolite 4 A. Figures 72 and 73 provide additional information regarding the areas in the United States and Western Europe in which zeolite 4 A has had a major impact.

The replacement of phosphate by zeolite 4 A has gained additional impetus as a result of changes in the prices of raw materials.

The economic situation with respect to other detergent ingredients is highly variable from continent to continent. Moreover, these materials are used in a much smaller quantity; thus, their economic impact is less significant and will not be explored further here.

9.2 Detergents

In 1982, total world production of soaps, detergents, and cleansing agents was $>30 \times 10^6$ t. Again, the tonnage is variable among the continents and individual countries (Table 44). Table 45 provides an example of the way detergent production

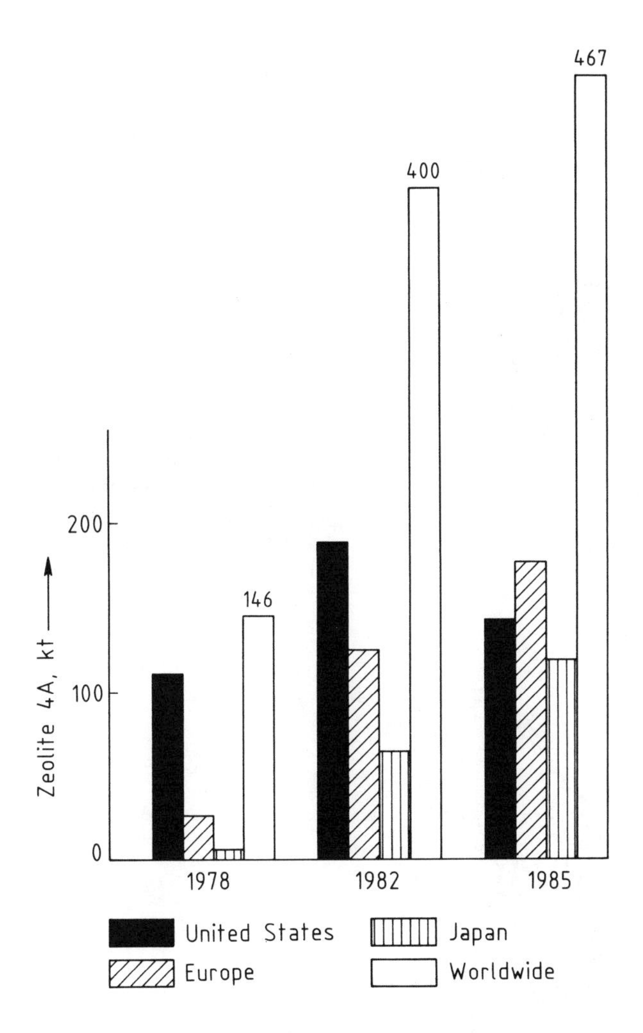

Figure 71. World production capacities of zeolite 4 A [482]

has developed quantitatively in comparison to fabric softeners, dishwashing agents, and household cleansers (data from the Federal Republic of Germany for the years 1975–1982).

Table 46 shows the change in overall production of soaps, detergents, and cleansing agents and their percent distribution worldwide from 1960 to 1982. Yearly growth rates fell at a relatively constant rate from 6.5% to 3.7% between 1972 and 1982 (Table 47).

Information regarding the consumption of soaps, detergents, and cleansing agents in various Western European countries for 1982 is found in Table 48. Data are included for per capita soap and detergent consumption.

Worldwide per capita use of synthetic detergents and cleansing agents has varied greatly (Table 40).

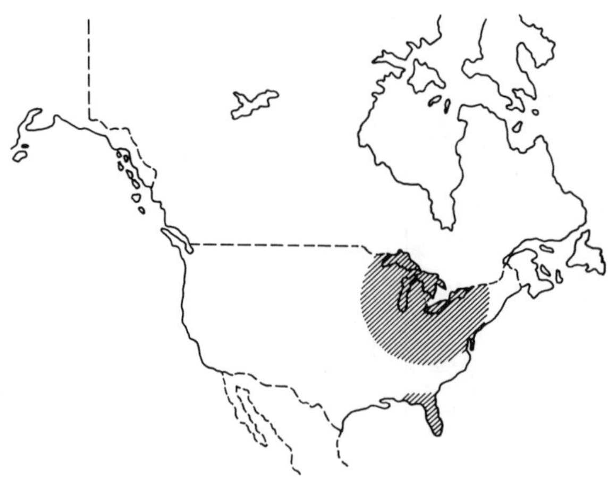

Figure 72. Primary distribution areas in North America for detergents containing zeolite 4 A [486]

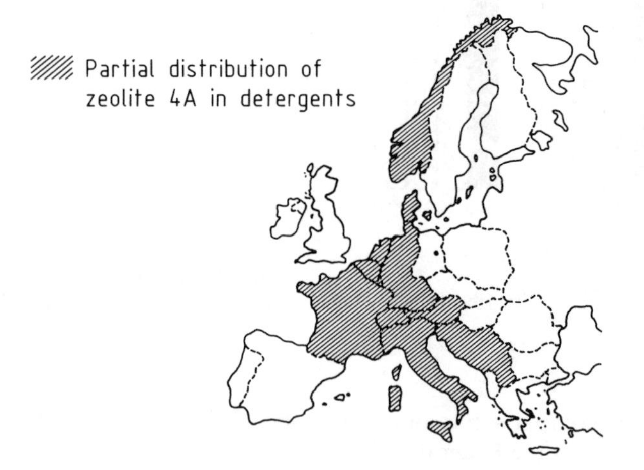

Figure 73. Primary distribution areas in Europe for detergents containing zeolite 4 A [486]

Table 44. World production of soaps, detergents, and cleaning compounds in 1982, in 1000 t

Country	Soaps	Detergents[a]	Scouring cleaners	Other cleaning compounds[b]	Total
Austria	3	81	10	20	114
Belgium–Luxembourg	28	229	16	45	318
Denmark	8	156	1	27	192
Federal Republic of Germany	134	1126	34	415	1709
Finland	13	43	1	4	61
France	110	820	42	125	1097
Greece	16	158			174
Iceland		3		1	4
Italy[c]	121	772	71	76	1040
Norway	9	29	1	4	43
Portugal	53	102	5	1	161
Spain	32	755	33	59	879
Sweden	13	126			139
Switzerland	10	106	4	21	141
The Netherlands	26	193	25	40	284
United Kingdom	168	914	61	142	1285
Yugoslavia	32	240	47	17	336
Western Europe total	*776*	*5853*	*351*	*997*	*7977*
German Democratic Republic	31	249			280
Poland	79	244			323
USSR	1091	1076			2167
Others	145	443		8	596
Eastern Europe total	*1346*	*2012*		*8*	*3366*
Canada	29	245		30	304
United States	545	5196	529	1415	7685
North America total	*574*	*5441*	*529*	*1445*	*7989*
Argentina	110	116	4	5	235
Brazil	660	355		17	1032
Mexico	249	653	15	28	945
Others	413	363		3	779
Latin America total	*1432*	*1487*	*19*	*53*	*2991*
Oceania total	*66*	*268*	*9*	*14*	*357*
Africa total	*976*	*569*	*10*	*42*	*1597*
China	1038	569			1607
India	940	169			1109
Japan	145	922			1067
Others	1114	899		2	2015
Asia total	*3237*	*2559*		*2*	*5798*
World total	*8407*	*18 189*	*918*	*2561*	*30 075*

[a] Including household cleaners and dishwashing agents. [b] In European countries including fabric softeners, in the United States moreover bleaching agents. [c] Consumption figures.

Table 45. Production of synthetic detergents and cleansers* in the Federal Republic of Germany (1000 t) [481]

Products	1975	1976	1977	1978	1979	1980	1982
Detergents for textiles (incl. auxiliary agents)	652.7	715.0	679.4	711.1	738.7	766.2	783.1
Fabric softeners	274.1	312.2	335.4	379.5	401.1	413.4	410.7
Dishwashing detergents	160.8	165.5	185.3	202.1	224.6	212.1	215.3
Household cleansers and scouring agents	131.3	136.8	124.7	146.1	169.8	175.6	166.5
Total	1218.9	1329.5	1324.8	1438.8	1534.2	1567.3	1575.6

* Household and commercial use (excluding special industrial cleansers).

Table 46. Worldwide total production of soaps, detergents, and cleaning agents and their distribution

Products	1960		1968		1976		1982	
	1000 t	%	1000 t	%	1000 t	%	1000 t	%
Soaps	6886	63.0	6493	42.0	8267	34.0	8705	28.9
Detergents	3433	31.5	7370	47.5	13547	56.0	17890	59.5
Scouring agents	461	4.0	585	4.0	626	2.5	917	3.1
Others	139	1.5	1015	6.5	1808	7.5	2562	8.5
Total	10919	100.0	15463	100.0	24248	100.0	30074	100.0

Table 47. World growth rates of soaps, detergents, and cleaning agents

Period	Average growth rate per year, %
1970–1972	6.5
1972–1974	5.1
1974–1976	5.1
1976–1978	5.1
1978–1980	4.1
1980–1982	3.7

Table 48. Western European consumption of soaps and washing and cleaning compounds in 1982

Country	Products, 1000 t				Population (10^6)	Per capita consumption, kg		
	Soaps	Detergents	Scouring cleaners, softeners	Total		Soaps	Detergents cleaners, softeners	Total
Austria	9	92	30	131	7.3	1.2	16.8	18.0
Belgium–Luxembourg	34	214	12	260	10.2	3.3	22.2	25.5
Denmark	12	101	23	136	5.1	2.3	24.2	26.5
Federal Republic of Germany	116	1073	440	1629	61.6	1.9	24.5	26.4
Finland	7	42	5	54	4.8	1.5	9.7	11.2
France	97	904	169	1170	54.3	1.8	19.8	21.6
Greece	17	159		176	9.8	1.7	16.2	17.9
Iceland	2	3		5	0.2	0.8	12.9	13.7
Ireland	3	35	3	41	3.5	0.9	10.9	11.8
Italy	121	772	147	1040	56.4	2.2	16.3	18.5
Norway	10	56	3	69	4.1	2.4	14.4	16.8
Portugal	37	102	6	145	9.9	3.7	11.0	14.7
Spain	20	695	92	807	37.8	0.8	21.8	22.6
Sweden	17	155	1	173	8.4	2.0	18.7	20.7
Switzerland	11	106	25	142	6.4	1.8	20.5	22.3
The Netherlands	20	202	51	273	14.3	1.4	17.7	19.1
United Kingdom	109	818	203	1130	56.1	1.9	18.2	20.1
Yugoslavia	25	220	74	319	22.4	1.1	13.1	14.2
Western Europe total	667	5749	1284	7700	372.6	1.8	18.9	20.7

9.3 Fabric Softeners

Europe is by far the major consumer of fabric softeners. A number of reasons explain this, including the fact that European water is relatively hard and the fact that drum-type washing machines equipped with separate dispensers for auxiliary agents are very popular in Europe. By contrast, products in this class continue to be relatively insignificant in the remainder of the world. Figure 74 provides information about European fabric softener production for the year 1982. In these production figures, the majority of the products are simple fabric softeners with an active ingredient content of ca. 5%. Figure 75 shows the growth of fabric softener production and per capita consumption of such products in a few Western European countries from 1975 to 1981. Consumption rose particularly rapidly between 1975 and 1979. The Federal Republic of Germany is the clear leader in European fabric softener consumption, with a per capita use of 6.4 kg in 1982.

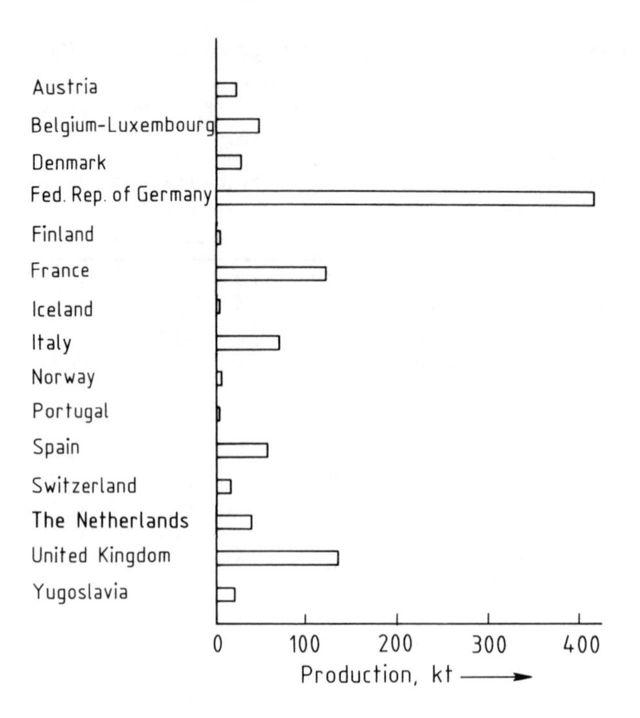

Figure 74. Production of fabric softeners in Europe in 1982 [487]

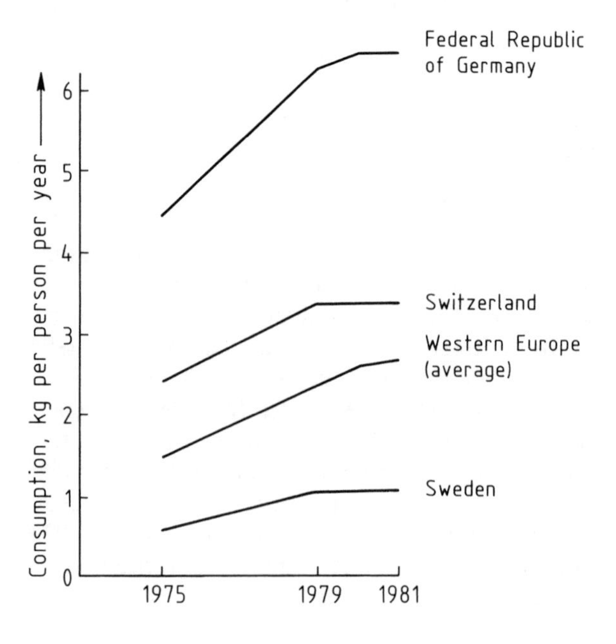

Figure 75. Consumption of fabric softeners in various countries (total: 900 000 t) [488]

10 Ecology

10.1 Laundry, Wastewater, and the Environment

The washing process represents a complex interaction between soiled laundry, water, both mechanical and heat energy, and detergents. Laundry is repeatedly cycled through the system, and clean laundry can be regarded as the product. Wastewater is the troublesome byproduct, which has the potential for causing a number of undesirable effects in sewage treatment plants as well as in the environment.

Virtually all of the clean water brought into the process is later released to the sewage system in the form of highly contaminated wastewater, water containing extra energy (heat), the soil from the laundry (lint, dyes, finishing agents, etc.), and detergents.

Detergent components are released to the wastewater either in essentially unchanged form or else as the products of reaction with other materials present. The principal contributions of detergent formulations to wastewater are surfactants and phosphates. Therefore, surfactants and phosphates—and the ecological and legal consequences with which these have become associated—are the subject of detailed consideration later in this section. The remaining detergent components are much less significant in this context because of their very low concentrations, their biodegradability, or their apparent or proven harmlessness in the environment.

10.2 General Criteria and Assumptions Regarding the Treatment of Laundry Wastewater

Laundry wastewaters vary considerably in concentration and composition. These differences arise partly due to variations in laundry soil levels, although washing technology and the composition and amount of added detergent are also significant factors. The differences are particularly great between the wastewater generated by household laundry and that from commercial operations. The latter generally contains lower concentrations of contaminants originating from detergents, largely as a result of more economical use of specialty detergents and the use of water-softening equipment.

Little doubt exists that laundry wastewater must be generally regarded as a heavily contaminated medium; it cannot be returned to receiving waters in untreated form.

Fortunately, the wastewater emanating from the millions of household washing machines is widely distributed, and in a pattern consistent with that of general water consumption. As a result of dilution in the public sewerage system and in sewage treatment plants, both the temperature and the relatively high pH of wastewater are moderated considerably, and the effect of peak loads is also somewhat compensated. Only for this reason is biological treatment of laundry wastewater in a normal sewage treatment plant even feasible. Otherwise, major problems would be raised in dealing with the load of organic pollutants introduced by household and commercial laundry operations. Such dilution within a typical public sewage system normally exceeds a factor of 10.

10.3 Contribution of Laundry to the Sewage Load [489], [490]

Table 49 presents calculated estimates of median theoretical concentrations of the most important detergent components in public wastewater. These estimates are derived from statistical data collected in the Federal Republic of Germany, and depend on both detergent consumption and municipal sewage volumes.

From the practical standpoint of successful management of a sewage treatment operation, the most important values are the calculated pollutant levels in relation to combined household and commercial water consumption (200 L per inhabitant and day). These have been based on the fact that ca. 70% of total surfactant production is consumed for laundry detergents and other cleansing agents. Actual values for concentrations and loads show considerable variation. Time of day is a factor, as is the day of the week (with Saturdays and Mondays producing the highest values). Even seasonal variations are apparent, but all of these fluctuations are reasonable when examined in the light of known household laundry habits [491].

The household laundering operation must be regarded as a major consumer of domestic water. In the Federal Republic of Germany, for example, only 10% of the laundry is washed commercially [492]. The amount of water consumed for this purpose can depend highly on the wash technology employed. Indeed, the literature contains values ranging from $< 1\ m^3$ to ca. $7\ m^3$ for the amount of water consumed in processing 100 kg of dry laundry. According to SCHULZE-RETTMER [493]–[495], a reasonable average value is $2-4\ m^3/100$ kg, assuming 0.4 kg of laundry per resident and day. This figure has been rising recently; the trend toward lower washing

Table 49. Average surfactant concentrations in sewage in the Federal Republic of Germany

	Daily per capita consumption	Calculated average concentration in municipal sewage, mg/L
Average water use (household wastewater, excluding industry, rainwater, etc.)	200 L [a]	
Detergent consumption (1980: 750 000 t/a) [b]	33.3 g	167
Anionic surfactant consumption (1980: 151 000 t/a) [c]	6.71 g	33.5
(ca. 70% of the total in detergents) [d]		(23.5)
Nonionic surfactant consumption (1980: 91 700 t/a) [c]	4.07 g	20.3
(ca. 70% of the total in detergents) [d]		(14.2)
Cationic surfactant consumption (1980: 26 100 t/a) [c]	1.16 g	5.83
(ca. 90% of the total in softeners) [d]		
Phosphorus consumption in detergents (1980: 56 800 t/a) [c]	2.52 g	12.6 [e]

[a] P. Koppe, based on studies conducted by the Ruhrverband; communicated in 1981 at a working group of the Hauptausschuß Phosphate (Phosphate Commission).

[b] Wirtschaft und Statistik, Aug. 1981, Statistisches Bundesamt, FRG, Kohlhammer Verlag, Mainz: population 61 654 000 (March 1980).

[c] Hauptausschuß Phosphate und Wasser (report of the NTA Working Group): 1980: 225 000 t of triphosphate = 56 800 t P/a; i.e., a reduction from the 1975 value of 69 000 t P/a.

[d] Deutscher Ausschuß für Grenzflächenaktive Stoffe; communication in the Hauptausschuß Detergentien (Detergent Commission), December 15, 1981.

[e] Municipal sewage also contains approximately the same amount of phosphate of fecal origin.

temperatures has brought with it an increase in consumption for both water and detergent despite intensive efforts to introduce laundering techniques that are more water- and energy-efficient.

According to the Schulze–Rettmer data [495], the per capita amount of laundry wastewater generated daily is 8–16 L, which translates into 4–8% of the total hydraulic load reaching a sewage treatment plant. Krüssmann and Hloch report per capita values of up to 26 L/d, or 13% of the total sewage volume [496].

Table 50 shows the extent to which various waters are altered by soil and detergent chemicals in the course of the laundering process.

A particularly important parameter whose value affects most calculations related to sewage treatment, as well as laundry wastewater evaluation, is the amount of biodegradable organic matter present, usually reported as the BOD_5 (biochemical oxygen demand). Regardless of whether the measuring criterion is the BOD_5, the COD (chemical oxygen demand), or the TOC (total organic carbon), the amount of

Table 50. Extent of soil load carried by various waters

Parameter	Unit	Drinking water	Mixed wastewater derived from laundering	Municipal sewage
Transparency	cm	700	2	1–4
Temperature	°C	15	25 * (max. 40)	10–20
pH		7.3	8–10	6.5–8.5
COD	mg/L	3–(10)	900–1300	200–600
BOD_5	mg/L	1	400–1000	150–400
Total N	mg/L	5	15–65	40–80
Total P	mg/L	0.2	100–300 (200)*	2–40 (20)*
Anionic surfactants	mg/L	0	15–180 (75)*	5–35
Nonionic surfactants	mg/L	0	5–90 (37)*	2–25
Soaps	mg/L	0	5–20 (12)*	<2

* Average.

organic contamination originating from soil is doubled or tripled in institutional laundry wastewaters, compared to laundry wastewater from households.

The average per capita load of biodegradable organic matter in community sewage is estimated to be 60 g of BOD_5 per day. SCHULZE–RETTMER reports [497] that the average BOD_5 of the wastewater from 100 kg of household laundry is 2.74 kg of BOD_5 (1.95–3.44). An assumption of the per capita laundry of 0.4 kg/d implies a soil load of 8.6–11 g BOD_5 per resident and day, or 16% of the total population equivalent value. A 100-kg batch of laundry thus produces a soil load equivalent to a population of 40–50 persons.

Normally, more than half of the total organic load of laundry wastewater is derived from organic compounds in the detergent, primarily soap and biodegradable surfactants. Nevertheless, only one half or less of the biodegradable contaminants load comes from this source. The remaining half is due to laundry soil.

10.4 Detergent Laws

The data presented for the fraction of hydraulic and soil load contributed by laundering to municipal sewage show that detergents and laundry soil represent the largest single nonfecal burden on the system. Thus, it is understandable that detergents have represented an early target for closer investigation. Surfactants and phosphates in particular became the subject of specific laws and regulations because of the magnitude of their contribution and their apparent ecological impact.

One of the earliest pieces of detergent legislation was the German Detergent Law of 1961 [5]. With its specific requirement for a minimum level of biodegradability, this law subjected a fundamental product parameter to legal control for the first time. The law, along with subsequent directives from the European Community [498]–[500] and other countries (e.g., France, The Netherlands, Switzerland, and Austria), placed a strict requirement of $> 80\%$ biodegradability on all anionic, nonionic, cationic, and amphoteric surfactants. Nonetheless, the statute that followed (1962) [6] provided only for the control of anionic surfactants. This legal move successfully displaced the poorly biodegraded tetrapropylenebenzenesulfonate (TPS) from detergent formulations by the readily biodegraded linear alkylbenzenesulfonate (LAS). Thus, by 1965 foaming problems in German sewage treatment plants and surface waters had been eliminated.

This first detergent law was a model piece of legislation upon which further developments in the past 20 years have been based, including corresponding regulations and voluntary agreements in various parts of Western Europe, North America, Brazil, and Japan [501].

What might tentatively be seen as the final result of these advances is the German law of August 20, 1975, regulating the "environmental compatibility of laundry detergents and cleansers (Detergent Law)" [502] including its statutes. This legislation contains the most sweeping demands enacted to date and thereby has had the most striking effects.

10.4.1 General Ecological Goals and Requirements

According to § 1 (1) of the German Detergent Law, placement of detergents and cleansers on the market is permitted only if the absence of all avoidable deterioration in the quality of surface water can be ensured. Particular attention is devoted to the potential role of surface water in the drinking water supply and to problems related to the operation of sewage treatment plants. Implementation of this fundamental demand entailed establishing a number of broad requirements, some of which were part of the law itself. Other requirements were implied in the authorization that the law granted for subsequent statutes. The effect of the requirements can be seen even in the earliest stages of detergent formulation as manufacturers were directed to provide thorough and prompt notification regarding the composition of all products offered. Consumers, too, were involved, in the sense that they were to be provided all necessary information regarding legal use of detergent products (§ 1, Section 2: "Detergent and cleanser usage must be consistent with the law and with the objective of water quality preservation.") The general requirements and authorizations for implementation applied directly to the following areas:

1) Unambiguous product labeling, including qualitative declaration of all significant components present
2) Deposition of frame formulations for all products in the Federal Environmental Office [503]
3) Utilization only of biodegradable organic components, particularly surfactants and other organic compounds viewed as potential phosphate substitutes
4) Provision of directions for dosage of detergents on all packages, including variations applicable to water of differing hardness, where hardness is defined in terms of four ranges
5) Obligation on the part of local water supply authorities to publicize the hardness characteristics of their water
6) Limitations on permitted phosphate levels, including some cases of a total phosphate ban, provided that ecologically sound substitutes are available

Detergent products affected by the Detergent Law that have the potential for entering surface waters include: household laundry detergents; household cleansers; dishwashing agents; rinsing and other laundry aids; detergents for commercial laundries; industrial cleansers; cleansing agents employed in the leather, tanning, paper, and textile finishing industries that might be released in wastewater; and personal hygiene products, such as shampoos, bubble baths, oral care products, and cleansing cosmetics.

10.4.2 Legal Limitations on Anionic and Nonionic Surfactants

In addition to the general requirements (§ 3) for biodegradability or the removal of surfactants and other organic materials from laundry detergents and cleansers, a specific statute was developed in 1977 [504] which spelled out concretely "test methods for measurement of the biodegradability of synthetic anionic and nonionic surfactants in laundry detergents and cleansers." In the sense of this statute, anionic surfactants are determined as "methylene blue active substances" (MBAS). Nonionic surfactants are defined as substances that are bismuth-active (BiAS) after passage through cation- and anion-exchange columns. Prior separation of anionic and nonionic surfactants from a detergent is required before biological testing is conducted to exclude the disruptive effect of mutual interactions. The German Detergent Law requires a minimum of 80% biodegradability for the total separated anionic and/or nonionic surfactants present in a packaged detergent, reported as % MBAS and % BiAS loss. In other words, only primary or functional degradation is measured, which for a surfactant means disappearance of a specific analytical reaction. This corresponds to the loss of the ecologically significant surface activity.

10.4.3 Control and Test Procedures

Two biological test methods are mandated for establishing biodegradability:

1) OECD Screening Test
2) OECD Confirmatory Test

Detergents and/or anionic and nonionic surfactants are considered biodegradable if they pass the 80% degradation level in the screening test. If the degradation rate proves lower, or if doubt remains, a subsequent confirmatory test is required, and the results of this test are regarded as definitive. The latter test is the same as that specified as the referee test in the directives of the European Community, although these directives also recognize, for screening purposes, the French AFNOR Test and the British Porous Pot Test in addition to the OECD screening procedure [499], [500], [505], [506].

The OECD Screening Test is based on a static shake flask method and corresponds to surface water conditions. In this procedure, a mineral salt solution is incubated with 5 mg/L of MBAS or BiAS as the sole source of carbon. The MBAS or BiAS loss is determined at fixed intervals for up to 19 d, and the results are compared to the behavior of two control substances: the poorly degradable TPS (< 35% loss) and the readily degradable LAS (92% loss). An example from such a determination is illustrated schematically in Figure 76. Biodegradability data — MBAS and BiAS removal as well as carbon removal and BOD — that are elaborated in screening tests with a number of surfactants are listed in Table 51.

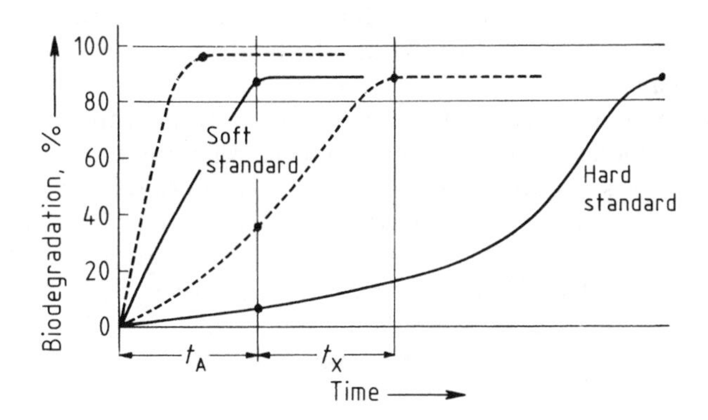

Figure 76. Biodegradability evaluation in the OECD Screening Test
t_A: maximum 14 days; $t_A + t_X$: maximum 19 days

Table 51. Surfactant biodegradation in screening tests

Surfactants	Primary biodegradation, OECD Screening Test, % MBAS/BiAS removal	Ultimate biodegradation in	
		Closed Bottle Test, % ThOD*	Modified Screening Test, % C removal
Anionic surfactants			
LAS	95	65	73
TPS	8–25	0–8	10–13
C_{14-18} α-Olefinsulfonates	99	85	85
C_{13-18} *sec-* Alkanesulfonates	96	73	80
C_{16-18} Fatty alcohol sulfates	99	91	88
C_{12-15} Oxo alcohol sulfates	99	86	
C_{12-14} Fatty alcohol diethylene glycol ether sulfates	98	100	
C_{16-18} α-Sulfo fatty acid methyl esters	99	76	
Nonionic surfactants			
C_{16-18} Fatty alcohols 14 EO	99	86	80
C_{12-14} Fatty alcohols 30 EO	99	27	
C_{12-14} Fatty alcohols 50 EO	98		
C_{12-18} Fatty alcohols 6 EO 2 PO	95	83	69
C_{12-18} Fatty alcohols 5 EO 8 PO	70	15	
C_{12-14} Fatty alcohols 10 PO	50–63	21	11
C_{13-15} Oxo alcohols 7 EO	93	62	
Isononylphenol 9 EO	6–78	5–10	8–17
C_{8-10} *n*-Alkylphenols 9 EO	84	29	
C_{12-18} Amines 12 EO	88	33	
EO/PO Block polymers	32	0–10	18

* ThOD = Theoretical oxygen demand.

The OECD Confirmatory Test is a continuous procedure more closely related to practice. It simulates through a prescribed test procedure the biodegradation that occurs in an activated sludge plant (Fig. 77).

In this procedure, a solution of MBAS (20 mg/L) or BiAS (10 mg/L) in synthetic sewage is continuously fed into a model plant, resulting in a mean retention time of 3 h. After inoculation of the test system according to a prescribed procedure and growth of the activated sludge, a plateau in the degradation rate is eventually reached with readily degradable substances. This initial phase is followed by a 21-d evaluation period, using a minimum of 14 daily values based on samples from 24-h collection periods. From these data an average degradation rate is calculated. Surfactants that are degraded only with difficulty produce a lower, irregular curve which fails to show a distinct plateau (Fig. 78).

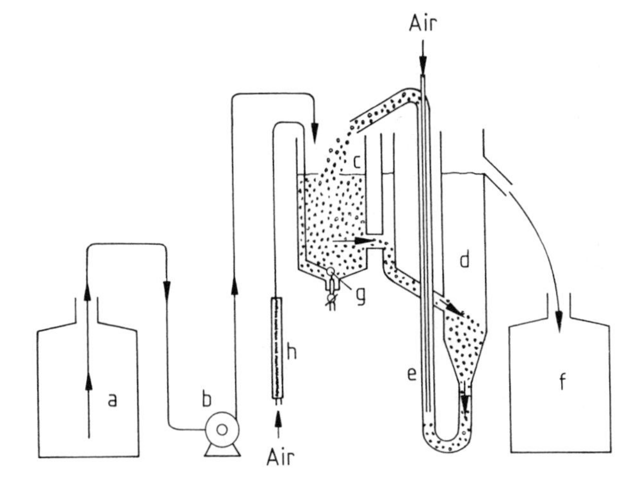

Figure 77. Experimental arrangement for the OECD Confirmatory Test as specified by the German Detergent Law
a) Sample container;
b) Dosage pump; c) Activated sludge vessel (capacity, 3 L);
d) Settling vessel; e) Air lift;
f) Collection vessel; g) Fritted disk; h) Air flow meter

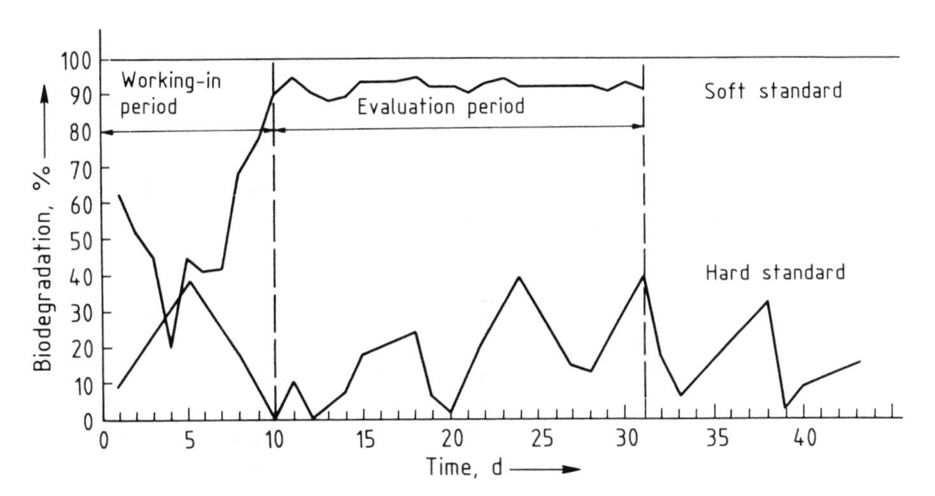

Figure 78. Determination of a rate of biodegradation according to the method specified in the OECD Confirmatory Test

In the Coupled Units Test, two such units are run in parallel, whereby one is fed with synthetic sewage plus test compound, e.g., in a concentration corresponding to 20 mg of carbon per liter, while the blank unit is fed only with synthetic sewage. If the effluents differ, e.g., by 5 mg of carbon per liter, then the test compound has reached 75% carbon removal. A schematic presentation of the Coupled Units Test is given in Fig. 79.

Surfactant biodegradability data elaborated in the OECD Confirmatory Test and the Coupled Units Test are presented in Table 52.

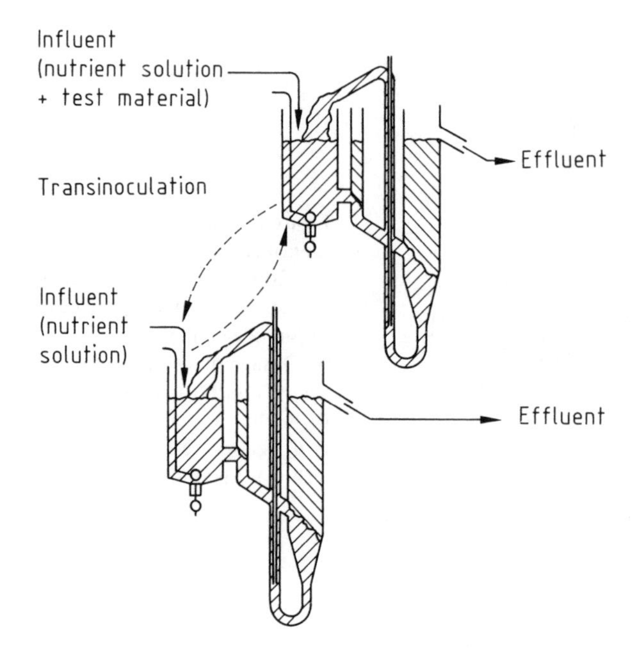

Figure 79. Modification of the OECD Confirmatory Test (Coupled Units Test)

10.4.4 Ecological Characteristics of the Major Surfactants and Practical Effects of the Detergent Law

One practical effect of detergent legislation has been the disappearance in the Federal Republic of Germany, Europe, the United States, and Japan of essentially all of the significant ecological problems caused by anionic surfactants, and this has been a direct consequence of a single major change: replacement of TPS by LAS. Other progress includes the acquisition of a detailed picture of further steps in the biodegradation process of LAS, including that for aromatic ring structures on the way to their mineralization. Insight has also been gained into degradation mechanisms and pathways. Equally important, or perhaps even more important, has been the knowledge gained for a wide variety of other available anionic surfactants, including the olefin- and alkanesulfonates, α-sulfo fatty acid methyl esters, derivatives of modestly branched synthetic alcohols, and the readily biodegradable fatty alcohol sulfates, all of which show greater than 80% MBAS loss and extensive mineralization. In addition, a large number of nonionic surfactants now exist with BiAS removal > 80%. Primary and ultimate biodegradability of these materials has been shown to increase in the following order for ethoxylates: nonylphenol < secondary alcohols < oxo alcohols < fatty alcohols. Nevertheless, the great variety and heterogeneity of nonionic surfactant structural types has made apparent the subtle differences in size and structure of hydrophobic and hydrophilic groups, which can have a major effect on biodegradability, a phenomenon that is less impor-

Table 52. Surfactant biodegradation in sewage treatment plant models

Surfactants	Primary biodegradation, OECD Confirmatory Test, % MBAS/BiAS/DAS* removal	Ultimate biodegradation, Coupled Units Test, % C/COD** removal
Anionic surfactants		
LAS	90–95	73 ± 6 (C)
TPS	36	41 ± 9 (COD)
C_{12} Fatty alcohol sulfate	99	97 ± 7 (C)
C_{13-18} *sec*-Alkanesulfonates		93 ± 5 (C)
C_{16-18} α-Sulfo fatty acid methyl esters		98 ± 6 (C)
Nonionic surfactants		
C_{16-18} Fatty alcohols 10 EO	98	62 ±28 (C, 3 h) 90 ±16 (C, 6 h)
C_{12-14} Fatty alcohols 30 EO	98	59 ±20 (C)
C_{11-15} *sec*-Alcohols 9 EO	86	36 ± 9 (C)
C_{11-15} *sec*-Alcohols 8 EO 5 PO		24 ± 5 (C)
C_{13-15} Oxo alcohols 12 EO	96	59 ± 6 (C)
Isononylphenol 9 EO	97	48 ± 6 (C)
C_{8-10} *n*-Alkylphenols 9 EO	96	68 ± 3 (C)
EO/PO Block polymers	7	2 ± 4 (C)
Cationic surfactants		
Cetyltrimethylammonium bromide (CTAB)	98	104 ± 6 (C, 6 h)
Dodecylbenzyldimethylammonium chloride (DBDMAC)	96	83 ± 7 (C)
Distearyldimethylammonium chloride (DSDMAC)	94	108 ± 9 (C)

* DAS = Disulfine blue active substance. ** C/COD = Carbon or chemical oxygen demand.

tant with anionic surfactants. Ultimate biodegradation pathways have recently become known for most of the nonionic surfactants as well.

Years of systematic monitoring of sewage treatment plants and rivers has shown that, in general, the residual concentration of surfactants in streams is extremely small. Indeed, these values have shown a consistent decrease in recent years despite increasing use of surfactants, a phenomenon that can be explained by the growth in sewage treatment capacity. Even in a river as heavily burdened as the Rhine, annual average MBAS concentrations have, with few exceptions, remained in the range of 0.02–0.1 mg of MBAS per liter, although changing patterns of use have caused BiAS residual concentrations to approach levels equivalent to the MBAS concentrations. Illustrative of the current situation is the fact that the Rhine, where it crosses the Germany–Netherlands border (i.e., after serving the needs of a region with ca. 40×10^6 inhabitants), is only transporting 0.68 % and 1.59 %, respectively, of the total quantities of anionic and nonionic surfactants consumed by the relevant popu-

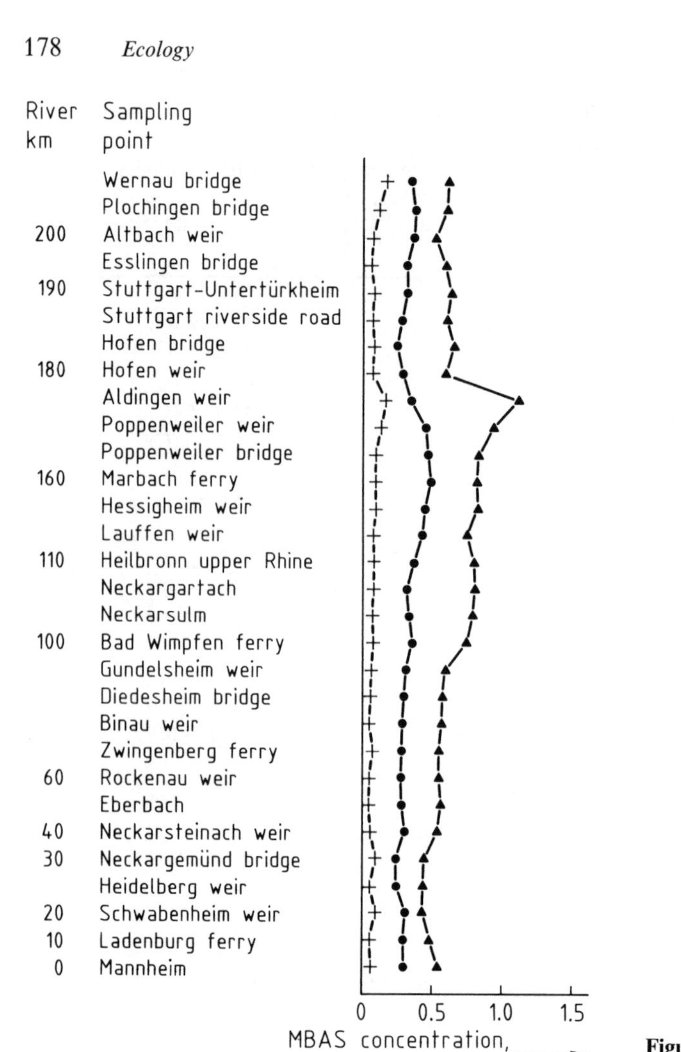

Figure 80. Anionic surfactant concentrations in the Neckar (annual means 1964, 1965, and 1983)

lation [507]–[510]. The Neckar gives an illustrative example of surfactant concentration developments in German rivers (Fig. 80) [510a]. The Detergent Law (§ 1) calls unambiguously for total ecotoxicological safety with respect to detergents, but specific provisions with respect to measuring fish toxicity, for example, have not been established. It is safe to assume, however, that the highly biodegradable anionic and nonionic surfactants allowed by the law are of only marginal toxicity to fish. Even so, studies of surfactant fish toxicity have become customary, and these studies have often shown that toxicity is inversely proportionate to biodegradability. Some acute toxicity data of various surfactants are listed in Table 53.

Table 53. Acute toxicity of surfactants

Surfactants	LC_{50} (fishes), mg/L	LC_{50} (daphniae), mg/L	NOEC* (algae, growth inhibition), mg/L
$C_{11,6}$ LAS	3–10	8–20	30–300
C_{14-18} α-Olefinsulfonates	2–20	5–50	10–100
Fatty alcohol sulfates	3–20	5–70	60
Alcohol ether sulfates	1.4–20	1–50	65
Alkanesulfonates			
C_{13-15} up to $C_{16,3}$	2–10	4–250	
C_{15-18} and C_{18}	1–2	0.7–6	
Soaps			
0 °d	6.7		
5 °d	20–150		10–50
Fatty alcohol poly(ethylene glycol) ethers			
C_{9-11} to C_{14-15}			
2–10 EO	0.25–4	2–10	4–50
10 EO	1–40	4–20	
C_{16-18}			
2–4 EO	100	20–100	
5–7 EO	3–30	5–200	
10–14 EO	1.7–3	4–60	
Nonylphenol poly(ethylene glycol) ethers			
2–11 EO	2–11	4–50	20–50
20–30 EO	50–100		
EO/PO Block polymers	100	100	
Fatty alcohol EO/PO adducts (> 80% biodegradable)	0.5–1	0.3–1	
Distearyldimethylammonium chloride	1.5–40	4–100	

* NOEC = No observed effect concentration.

10.4.5 Regulation of Maximum Phosphate Concentrations

In itself, phosphate is not harmful. Indeed, it is an essential macronutrient for plants, animals, and humans. Normally, the phosphate concentration in surface waters is so low that it is a limiting factor for biological growth. Consequently, when excess phosphate is released into surface waters, the result is overfertilization, leading to increased algal growth. Biodegradation of the algae in stagnant surface waters leads to oxygen depletion in the lower water layers, which in turn causes a general reduction of overall water quality. The result may be loss of potential drinking water and, in some cases, damage to fishing interests. For this reason, use of sodium triphosphate in detergents has come under increasingly critical scrutiny. Detergent phosphates released with laundry wastewater are very rapidly converted into orthophosphate. Nevertheless, detergent phosphates represent only one of many factors contributing to eutrophication in surface waters.

A thorough study on phosphates entitled "Phosphorus: Pathways and Fate in the Federal Republic of Germany" was commissioned by the German Federal Ministry of the Interior and carried out by the Phosphate Commission within the Water Chemistry Section of the Gesellschaft Deutscher Chemiker (GDCh, German Chemical Society) [511]. More than 90% of the phosphate produced in or imported into the Federal Republic of Germany in 1975 (807 000 t of P/a) was found to be used in the fertilizer and feed industries, and only 69 000 t of P/a (ca. 8.6%) was consumed by the detergent and cleanser market. By 1985 this had declined further to 35 000 t as a result of the continuing replacement of phosphate by substitutes.

The same phosphate study led to the conclusion, however, that apart from some local variation, ca. 60% of the phosphate encountered in municipal sewage originated from detergents or cleansers. As a consequence of the partial removal of phosphate by sewage treatment plants and the input of phosphates from other sources, the share of detergent phosphate in surface waters was estimated to be ca. 40% (ca. 40 000 t of P/a). This phosphorus balance shows that the removal of phosphates from detergents alone cannot possibly solve the entire problem of surface water eutrophication.

The most comprehensive approach involves chemical elimination of phosphate at the sewage treatment plant, since this also permits removal of the large amount of phosphate of fecal origin. Precipitation techniques for phosphorus removal have already been introduced in both Sweden and Switzerland, the sewage being treated prior to discharge to stagnant water bodies. Phosphorus removal by this means is also being practiced in a growing number of sewage treatment plants in the Federal Republic of Germany [512].

Widespread application of such tertiary sewage treatment is regarded as an appropriate long-term goal, but German lawmakers still felt immediate action was necessary, thus imposing specific regulatory controls on the use of phosphates in detergents. The Detergent Law had already required that after 1975, detergent users be provided with dosage instructions based on water hardness data. In addition, the legislation limited maximum allowable phosphate contents in detergents and cleaners in a stepwise fashion with the phosphate regulation of June 5, 1980 [513]. The first stage went into effect on October 1, 1981, with the second stage following in January, 1984. As an illustration of the magnitudes involved, for water having a hardness between 0 and 7 on the German scale (equivalent to 0–125 ppm $CaCO_3$/L, i.e., relatively soft water), the permissible level of phosphate application in home laundry detergents was reduced from 2.8 g/L to 2.0 g/L. A higher level is permitted in areas where the water is harder. Rather than providing different detergent formulations for different localities, most manufacturers have responded by maintaining a single formulation but also providing consumers with a chart showing the precise amount of detergent to be used for a load of laundry in water of a given degree of hardness. Thus, the homemaker is left with the responsibility of deciding whether or not to comply with the law.

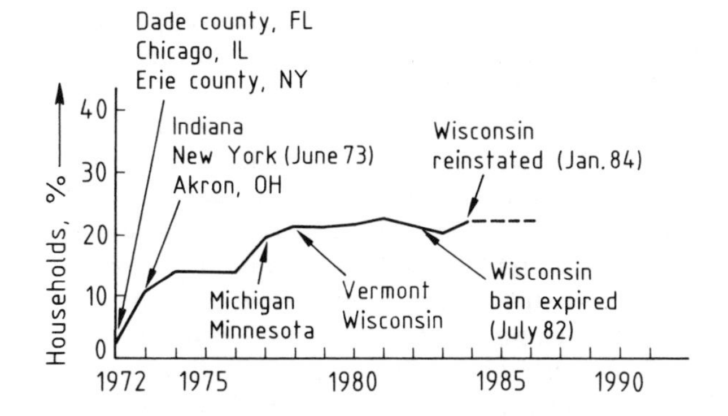

Figure 81. History of phosphate bans in the United States [42]

Laws and voluntary agreements with respect to phosphate use exist in other nations as well. Some are comprehensive in nature, whereas others apply on a regional basis. The history of legislation banning phosphate use in the *United States* is summarized in Figure 81. Currently such bans cover areas representing ca. 20% of all households, down slightly from a figure of 22% in 1981. Currently, the use of phosphate in home laundry detergents is banned entirely in six states: Indiana, Michigan, Minnesota, New York, Vermont, and Wisconsin. In addition, a number of communities have imposed their own bans. A total ban is imposed, for example, in Dade county, Florida, which includes the city of Miami. The same is true of both Chicago, Illinois, and Akron, Ohio, along with some of their suburbs. Other localities have chosen to impose limits. The states of Connecticut, Florida, and Maine, as well as ca. 15 municipalities, have set a limit of 8.7% phosphorus content in household laundry detergents.

In *Canada* a federal limit, established as a phosphorus content of 2.2%, has been placed on household detergent use of phosphates.

The problem of phosphate-induced eutrophication in Europe has been treated as a regional or national issue rather than as a pan-European matter requiring action by the European Economic Community. Thus, national or regional authorities have been given the responsibility for finding their own solutions.

Most other countries that have perceived the need for action have elected to impose limits on phosphate use. The usual limits involve complex formulas that depend on factors such as water hardness.

The government in *Italy* introduced a proposal for phosphate limits in March 1982, as one feature of a general environmental protection law. The Senate then adopted this government-proposed legislation in a form calling for a maximum allowed phosphorus content in detergents after June 30, 1987 of only 1%. This limitation applies without distinction to all laundry and dishwashing detergents and cleansers, irrespective of whether they are intended for household, commercial, or industrial use. Final approval by Parliament is still pending. In *The Netherlands,* industry joined together in a voluntary agreement to reduce phosphate levels in two stages; the second of these went into effect on January 1, 1983.

Some countries, notably *Sweden*, have concluded that the best approach is through tertiary sewage treatment, and by the end of 1983, Sweden had constructed 850 such plants serving nearly 80% of the Swedish population.

Switzerland has perhaps the most severe eutrophication problems in Europe as a result of its unique geology. The nation's legislators began considering detergent phosphate regulations in 1977, but not until December 8, 1980 did the federal government approve a plan calling for a two-stage reduction in permissible phosphate limits. The first stage went into effect on October 1, 1981, and the second on January 1, 1983. The rules were complex, and a range of phosphate concentrations was permitted, depending on individual circumstances. The Swiss Federal Parliament later amended the existing detergent legislation (on July 3, 1985) with the result that all use of phosphates in laundry detergents was banned after July 1, 1986.

Norway is also beset with special problems, especially with respect to Lake Mjøsa. Therefore, the Norwegian government has decreed that after January 1986, no household laundry detergent is permitted to contain more than 12% sodium triphosphate.

Phosphate limitation is currently under discussion in *Austria,* but no action has yet been taken. In *Finland* a "gentlemen's agreement" is in effect between industry and government, calling for a phosphate limit equivalent to a maximum phosphorus content of 7%.

Australia has also been the scene of discussions on the subject of phosphate limits.

Certain of the prefectures in *Japan* have instituted total bans on phosphate use. In view of the marketing patterns that exist in the country, manufacturers have chosen to respond by undertaking total reformulation of their products. As a result, within a period of only 2 years, it was reported that 90% of all detergents produced and sold in Japan were free of phosphates, being formulated instead with zeolite [144], [486].

The detergent industry has over the course of many years invested considerable amounts of research time and effort worldwide in the search for new phosphate substitutes that could make a contribution to the elimination of the eutrophication problem. In particular, a wide variety of organic sequestering agents has been investigated, including both polymers and low molecular mass materials [514]–[516].

The first major breakthrough came with the development and introduction of inorganic, water-insoluble sodium aluminum silicates, especially zeolite 4A. This material, used in combination with triphosphate or other phosphate-free sequestering agents or ion exchangers, leads to systems with excellent detergency properties and a low phosphorus content. Formulations based on zeolite 4A ensure the thorough maintenance of the detergency standards to which the public has become accustomed. These formulations are also attractive from the standpoint of raw material supply and economics, but above all they are important in terms of the contribution they have made to preservation of the quality of surface waters [486], [517]–[519].

The human toxicology (Chap. 11) and the ecological characteristics of zeolite 4A have been the subject of major research programs since 1973 in Germany (Henkel

KGaA)—partially sponsored by the German government—and in the United States (Procter & Gamble). The comprehensive nature of this investigation is unprecedented for a detergent raw material, and it has involved laboratory testing, model studies, and a wide variety of field tests (sewage treatment plants, fish ponds, etc.) [520]–[522]. All of these investigations have led to the same conclusion: there is no basis for concern regarding the detergent use of zeolite 4 A.

The introduction of zeolite 4 A into detergents in the Federal Republic of Germany has been the principal factor leading to a reduction in the amount of phosphate in detergents from 260 000 t of sodium triphosphate in 1975 to 140 000 t in 1985. The prerequisites and the prospects for further reduction of the phosphate content in detergents through use of organic sequestering agents and soluble ion exchangers are at this point very limited [523], [524]. In particular, the potential sequestering agent nitrilotriacetate is currently the subject of critical international discussion [486].

10.5 Other Detergent Ingredients

10.5.1 Cationic Surfactants

The use of cationic surfactants in the Federal Republic of Germany for detergency purposes during 1985 amounted to ca. 17 000 t.

The major contributors to this total were distearyldimethylammonium chloride (DSDMAC), used as a softening agent, and dialkylimidazolinium chloride (DAIC).

The earliest versions of international guidelines and national laws regarding detergent formulation and use made explicit reference to cationic and amphoteric surfactants, but no definitive test procedures for these materials were prescribed.

In the meantime, results have become available from various selective investigations, which provide a comprehensive overview of the environmental behavior of cationic surfactants [525].

When the usual degradation screening tests in mineral solution are conducted on cationic surfactants as the sole source of carbon present, virtually no degradation is observed. This result is partly due to the inhibitory action of the compounds themselves, which are biocides. Results obtained from activated sludge studies tend to be ambiguous, since distinction between true degradation and an elimination occurring through sorptive processes is difficult. Nonetheless, sewage treatment plant model studies using the OECD Confirmatory Test and the Coupled Units Test have shown high elimination rates for quaternary ammonium compounds, especially DSDMAC, and it was possible to ascribe a large fraction of this elimination to genuine biodegradation [526], [527]. Major differences in the biodegradability of various cationic surfactants apparently exist. Certain nonaromatic cationic surfactants have been found readily biodegradable [528]. Experiments involving ^{14}C-labeled compounds have shown the release of $^{14}CO_2$, demonstrating unambiguously that these compounds can undergo not only primary biodegradation, but also ultimate biodegradation. In addition, HUBER cites unpublished work by LARSON showing that the same is true for DAIC as well [529].

Investigations at selected municipal sewage treatment plants in the Federal Republic of Germany, Great Britain, and Belgium provide confirmation for the notion that > 90% elimination occurs under steady-state conditions. It seems unlikely that long-term elimination of this magnitude could be explained solely on the basis of adsorptive phenomena [530]–[532].

Toxicity with respect to aquatic organisms is an important criterion for judging ecological behavior [529]. Laundry softeners on basis of cationic surfactants exhibit LC_{50} values for fish in the range of 1.8–2.6 mg/L and are, therefore, relatively toxic. However, these acute toxicity data should not be applied directly to surface waters, since both sewage and surface waters contain an excess of anionic over cationic surfactants. This means that especially the poorly soluble fabric softening agents would be incapable of expressing their toxic properties due to their tendency to form water-insoluble neutral salts. In any case, the measured residual concentrations in receiving waters are extremely small due to the high rates of elimination observed in sewage treatment plants. Currently, these fall in the range of 5–30 μg/L. The corresponding safety factor in terms of acute LC_{50} values is at least 60. Bioaccumulation studies with ^{14}C-labeled DSDMAC in both drinking water and stream water, at the low concentrations relevant to the environment (20 μg/L), resulted in very low bioconcentration factors in whole fish (BCF 32 and 13, respectively) and in their edible portions (BCF 1).

The low residual concentrations found in streams and the low bioaccumulation potential suggest that risks due to cationic surfactant residues are insignificant.

10.5.2 Water-Soluble Inorganic Builders

Sodium carbonate (soda) and sodium silicate (water glass) are both alkaline and, thus, have the ability to raise the pH of sewage, but this results in no ecological problems.

10.5.3 Bleaching Agents

European laundry detergents generally contain up to 25% sodium perborate as a bleaching agent. During the wash process and in the wastewater, this leads to the production of free H_2O_2. The latter is rapidly inactivated, but H_2O_2 even at concentrations much higher than those present in laundry wastewater shows no toxicity with respect to bacteria. Thus, biological sewage treatment is unaffected by its presence [533]. The only ecological factor that must be taken into account is the resulting sodium borate, and with it the increase in boron concentration. Most of the boron found in water is of anthropogenic origin. Thus, due to its high mobility, boron in wastewater is sometimes used as an indicator of anthropogenic pollution [534].

A number of broad systematic studies have been carried out on the boron concentrations in rivers and drinking water. The values found for the Federal Republic of Germany are very low [535]. Boron levels in drinking water are so low as to be negligible from a toxicological standpoint (< 0.25 mg of boron/L), although some mineral waters, for example, exhibit values ranging from a few mg/L to much higher concentrations; by comparison, some fruits and vegetables contain as much as 80–300 mg/kg. Similarly, boron concentrations in German streams are also very low (Rhine: 0.1–0.25 mg/L; Neckar, Main, and Ruhr: 0.2–0.5 mg/L; polluted rivers such as the Nette, Nidda, and Niers: 0.5–1 mg/L; and those classed highly polluted, such as the Emscher: 1–2 mg/L). In certain other countries, geological conditions are such that much higher boron concentrations are measured. The toxicity of boron with respect to fish is quite low: $LC_{50} > 300$ mg/L. Boron levels as low as 1–2 mg/L possess a specific phytotoxicity with respect to certain agricultural plants, such as fruit trees, tomatoes, and vineyard stock. Therefore, it is not advisable to recycle boron-containing wastewater for irrigation purposes [536].

It is difficult to estimate what fraction of the boron in surface waters is derived from detergents, partly due to the great regional variability of both natural and industrial factors. One source [537] cites 73% as the proportion of anthropogenic boron in the Neckar in 1978, whereas another [539] estimates it to be ca. 50% in the Rhine at times of low water flow. At higher flow rates, the natural boron load dominates to an increasing extent. The Ruhr River has a very low natural boron content, and as such, it has been possible to ascribe the additional amounts introduced through wastewater largely to detergent use [534]. The 1975 boron consumption in detergents in the Federal Republic of Germany was 9500 t [539].

In certain other countries, especially in North America, bleaching is accomplished by a means that in Germany is currently largely restricted to commercial laundries: use of chlorine, an alternative that lends itself to application at a low wash temperature. Sodium hypochlorite, a substance that decomposes to give active chlorine, is usually the active ingredient. Chlorine is most certainly toxic in surface waters. Nevertheless, it does not reach the sewage treatment plant in the form of free chlorine, but rather is rapidly destroyed while still in the sewage system by chemical reaction with materials such as ammonia and various organic sewage components, which are readily oxidized or chlorinated.

10.5.4 Fluorescent Whitening Agents

Although fluorescent whitening agents are adsorbed by fabrics during the laundering process, they are also introduced to some extent into the wastewater in the course of repeated washing. Simple laboratory screening tests show fluorescent whitening agents to be biodegraded only to a very limited degree. Analysis conducted under conditions more closely resembling those found in a two-stage sewage treatment plant, however, revealed elimination values up to 96%. These are best explained on the basis of adsorption on the sludge. Total elimination has been demonstrated in a system involving tertiary treatment. From these results, one may conclude that sewage treatment essentially ensures the absence of whitening agents in surface waters.

Careful studies of such highly polluted rivers as the Thames in London, the Rhône at Lyon, the Seine at Paris, and the Rhine both at Basel and at Düsseldorf showed concentrations of fluorescent whiteners below the limit of their detection (0.25 µg/L).

Extensive investigations have been carried out with algae, mussels, and fish, and these have shown very low toxicity (LC_{50} values > 100 mg/L), indicating a biological safety factor in rivers of 10^6-10^8 [540].

10.5.5 Soil Antiredeposition Agents

These are primarily water-soluble, high molecular mass cellulose ethers. Their biodegradation proceeds as slow as that of natural cellulose, however. Such compounds are completely non-toxic to fish, and no significant ecological problems are known to be associated with their use.

10.5.6 Foam Regulators

Soap, which presents no ecological problems, is one of the most important foam inhibitors used in detergents. Other foam regulators used in very small amounts include silicones, which are virtually insoluble in water and undergo little biodegradation. Present knowledge suggests that these regulators are largely eliminated by the sludge in sewage treatment plants [541].

10.5.7 Detergent Enzymes, Fragrances, and Dyes

Enzymes are high molecular mass proteins. These are quickly inactivated and biodegraded by sewage treatment facilities and in receiving waters. Thus, there is no ground for concern about their effects with respect to maintaining water quality or their general environmental behavior.

This also applies to the other ingredients found in trace amounts in detergents, such as those designed to confer pleasing fragrance and appearance (perfumes, dyes, etc.). These materials are either decomposed or adsorptively eliminated in the course of sewage treatment. Trace residues, which might escape removal and thus appear in the environment, are ecologically insignificant.

10.6 Outlook

The preceding discussion on ecological aspects of detergents highlights the attention currently directed toward such matters in the course of detergent formulation and application. This concern has come to play a significant role alongside more traditional motives related to production and efficacy. Much valuable experience has already been gained through the successful implementation of laws and controls governing use of such primary detergent components as anionic and nonionic surfactants and phosphates. Thus, it is reasonable to anticipate that benefit/risk analyses will play an ever increasing role in the selection of raw materials and active ingredients in the future [502], [542], [543].

11 Toxicology

Detergents are products designed for everyday use; thus, they are found in large quantity in nearly every household. The home is quite unlike the industrial workplace: one cannot assume that every user of a household product plays strict attention to warning labels or recommended safety precautions. For this reason, any product offered on the market must insofar as possible be formulated so that it represents no significant or foreseeable health hazard, regardless of whether or not the product is properly handled. This in turn means that even at the earliest stages of raw material selection, great value should be attached to chemicals known to be harmless to humans following any realistic type of exposure.

The following types of exposure must be taken into account in assessing the risk presented by detergents and their components:

skin contact (wash liquor, residues remaining on laundered items, or exposure during manufacture)

ingestion (accidents, especially with children, and trace residues in drinking water)

inhalation (in the course of manufacture or use of powdered detergents)

Extensive toxicological studies are required, and the results must be carefully weighed, taking into account the amount and stability of individual substances. Such studies should examine local effects (skin irritation, development of allergic reactions on contact, skin penetration), systemic effects (both acute and chronic), and potential hazards of a more subtle nature (mutagenicity, embryotoxicity, and carcinogenicity).

The present section begins by examining the toxicology of the main detergent constituents, after which the properties of complete detergent formulations are explored. The approach is analogous to that followed in practice by a manufacturer in the course of product development.

11.1 Detergent Ingredients

11.1.1 Surfactants

Most of the biological properties of surfactants can be understood in terms of interactions occurring between surfactant molecules and such fundamental biological structures as membranes, proteins, and enzymes. Contact between a surfactant and a membrane leads to changes in membrane permeability, and in extreme cases, can even result in membrane solubilization. The most obvious potential consequence is interference with material transport and, ultimately, cell damage. A further danger is the fact that surfactants can significantly affect the resorption of other chemicals.

Proteins form adsorption complexes with both anionic and cationic surfactants. Complex formation is often a consequence of polar interactions between the hydrophilic residue of a surfactant and charged sites on the protein molecule, but hydrophobic interactions can also play a role. Such complex formation results in protein denaturation, which in the case of an enzyme implies a reduction or even total loss of catalytic activity, corresponding to a change in metabolic function [544].

Nonionic surfactants are characterized by their lack of strongly polar functionalities. Thus, compounds of this class rarely cause protein denaturation. They can induce a limited amount of protein solubilization, but the concentrations required for the appearance of harmful effects are usually greater than those of ionic surfactants.

Transport through skin is an important consideration with respect to detergent ingredients, which tends to be quite low for anionic and cationic surfactants [545], [546]. Transport is greater for nonionic surfactants, but again the amount of material capable of entering an organism by this route is so low that it can be essentially disregarded as a potential hazard [547]. By contrast, both anionic and nonionic surfactants can be readily resorbed through the gastrointestinal tract following their ingestion, whereas the intestinal resorption of cationic surfactants is low [548]. Even absorbed surfactants are relatively harmless because they are rapidly metabolized, mainly through β- or ω-oxidation of alkyl chains and some cleavage of poly-(alkylene glycol) ether linkages. Elimination occurs through the bile and the urine; significant accumulation within the body has never been demonstrated.

The ability of surface active agents to emulsify lipids means that repeated or prolonged exposure to surfactant-containing solutions can cause damage to the lipid film layer that covers the skin surface. As a consequence, the barrier function of the lipids is impaired, leading to increased permeability and loss of moisture. This is evidenced by dryness, roughness, and flaking of the skin. Very prolonged exposure to concentrated surfactant solutions can lead to serious damage and even necrosis. Skin tolerance varies widely among the compounds making up each class of surfactants. Nevertheless, one can generalize that tolerance to surfactants tends to increase in the order cationic, anionic, nonionic materials.

Broadly speaking, all of the commercially relevant surfactants are well-tolerated in the concentration ranges applicable to detergent use [549], [550]. However, certain cases of structure–activity relationships have been found. With nonionic surfactants (alkyl poly(alkylene glycol) ethers), for example, skin irritation diminishes with an increasing degree of ethoxylation [551]. For anionic surfactants, the length of the alkyl chain has been found to be closely related to skin irritability. Thus, within a given homologous series, compounds containing saturated alkyl chains of 10–12 carbon atoms show the strongest effects [552].

The eye is much more sensitive than skin to damage by small amounts of surfactants. Anionic surfactant solutions with a concentration above ca. 1 % can produce minor eye irritation, although this is normally reversible [552]. Serious damage is likely only if the eye comes in direct contact with a concentrated surfactant solution and if this contact is not followed by immediate and intensive flushing with water.

Virtually all surfactants that are employed in detergents have been the subject of extensive investigation with respect to their allergenic properties. In no case has an increased risk of allergy for the consumer been demonstrated. A few isolated examples of sudden, multiple allergic reactions in connection with anionic surfactants (alkyl ether sulfates or α-olefinsulfonates) were later traced to impurities of 1,3-sultones and chlorosultones, compounds that have long been recognized as potent allergens and that can arise by bleaching of surfactants with hypochlorite at low pH [553].

The acute oral toxicity of surfactants is low; LD_{50} values normally fall in the range of several hundred to several thousand milligrams per kilogram of body weight [554]. The major detrimental effect of surfactants is damage to the mucous membranes of the gastrointestinal tract. High doses lead to vomiting and diarrhea. If a surfactant reaches the circulatory system, it can cause damage even in very low concentration as a result of interactions with erythrocyte cell membranes, ultimately resulting in destruction of the cells (hemolysis). Inhalation of dust or aerosol containing high concentrations of surfactants can interfere with pulmonary functions [555]. This interference can be partly attributed to interactions with the surface active film of the vesicles of the lung [556].

The possibility of chronic toxicity has been a subject of intense investigation with members of all of the surfactant classes. Tests with experimental animals involving exposures for up to 2 years in dosage ranges of several thousand ppm have without exception shown complete safety [549], [550], [557]. Moreover, a number of investigations have been conducted over long periods of time with human volunteers. Substantial amounts of both anionic and nonionic surfactants have been administered, and no serious side effects have been discovered [554].

Neither long-term oral ingestion nor continuous skin exposure has ever suggested that surfactants within the three major groups possess carcinogenic activity [549], [558]. Similarly, surfactants can be regarded as completely devoid of both mutagenic [559] and teratogenic [549], [558] characteristics.

11.1.2 Builders

The most important substances in this category are sodium triphosphate, zeolite 4 A, and sodium nitrilotriacetate (NTA).

Pentasodium triphosphate is a compound that can be regarded as innocuous. Effects of triphosphate ingestion first become acute with dosages of several grams per kilogram of body weight, and even these manifestations are simply a result of localized damage caused by the high alkalinity of concentrated phosphate solutions [560].

Sodium aluminum silicate known as zeolite 4 A has undergone intensive toxicological investigation and has been found to possess absolutely no acute toxicity. The compound is also tolerated well locally. No adverse effects are seen with eye tissue beyond those associated with the presence of any foreign body. Studies on chronic toxicity or carcinogenicity following oral ingestion have also shown zeolite 4 A to be safe after long-term exposure. Inhalation of the material leads neither to silicosis nor to the development of tumors. Isolated reports of alleged carcinogenic activity have been associated with synthetic zeolites, but these have been shown to be related to the presence of natural zeolites possessing a fibrous morphology. The latter are indeed capable of inducing tumors, unlike zeolite 4 A, which consists of cubic crystals.

In summary, all studies agree in the conclusion that zeolite 4 A can be regarded as a fully inert substance with respect to living organisms. Zeolite 4 A appears to present no risks whatsoever either in the workplace or in home detergent use [561].

NTA shows little acute toxicity. This generalization is valid with respect to both inhalation and oral ingestion. Moreover, the compound produces minimal adverse effects on the skin even after repeated exposure, and it has no sensitizing properties. NTA is readily absorbed through the gastrointestinal tract, but it is then quickly eliminated without being metabolized. Numerous investigations support the conclusion that NTA is not a mutagen. Results from animal studies have shown that repeated applications of large amounts of NTA can cause changes in the urinary tract, including tumor formation, but these changes seem to be due to alterations in the metabolism of certain divalent cations by interaction with the complexing agent NTA. This effect is not considered relevant to conditions likely to be found when NTA is used in detergents. The possibility cannot be excluded that use of NTA could lead to trace amounts in drinking water. Nevertheless, concentrations to date have never been found to exceed a few micrograms per liter, and toxicological data suggest that this presents no human health risk [562]–[567].

11.1.3 Bleach-Active Compounds

Sodium perborate shows only a very slight acute toxicity. Extensive contact with concentrated solutions can cause irritation to the skin and mucous membranes, but this is a consequence of alkalinity rather than a specific effect of the substance itself. The rapid hydrolysis of perborate means that any physiological effects that are observed will be those of borates or boric acid. While it is true that these substances pass readily through the mucous membranes or through damaged skin surfaces, intact skin prevents excessive amounts from being absorbed. However, the intestinal tract is capable of admitting substantial amounts of boric acid, and the substance is considered lethal to adults at a dose of 18–20 g [568].

11.1.4 Auxiliary Agents

Enzymes. The proteolytic enzymes employed in detergents are quite safe from a toxicological standpoint. Evidence of local skin irritation appears to be rare, as are systematic effects following ingestion. As with all proteins, however, human allergenic reactions are possible. These reactions arise both through skin contact and through inhalation of enzyme-containing dust. Indeed, exposure to such dust in the early days of the manufacture of enzyme-containing detergents resulted in a number of asthmatic reactions among factory workers. This risk has been eliminated by drastic reduction in the enzyme concentration of industrial dust and by the introduction of enzyme prills.

The possibility that a consumer might inhale significant amounts of enzyme from detergent dust can be essentially excluded. Thus, the only plausible and relevant form of consumer exposure is skin contact in the course of dealing with handwashable items. At one time a definite relationship was assumed between skin reactions and use of enzyme-containing detergents, but a large body of evidence now shows the contrary. It appears safe to say that the presence of enzymes does not contribute to skin irritation by detergents, nor does it lead to increased risk of allergenic reactions [569], [570].

Fluorescent Whitening Agents. Much information is now available that supports the conclusion that the fluorescent whitening agents employed in detergents are safe materials. All studies on toxicity, teratogenicity, mutagenicity, and carcinogenicity are consistent in affirming that such compounds present no risk. Local tolerance of the substances is also good, and they display no tendency to cause sensitization. Moreover, long-term studies involving cutaneous application have failed to reveal any carcinogenic potential. Fluorescent whitening agents bind rather tightly to proteinaceous material; as a result, they are firmly retained by the upper keratin layer of the skin, ruling out any significant percutaneous transport. The correspondingly firm binding to textile fibers implies that risks of exposure by way of clothing can also be disregarded. In other words, all available toxicological and dermatological evidence supports the belief that low-concentration detergent use of fluorescent whitening agents presents no human health risk [571], [572].

11.2 Finished Detergents

Appropriate choice of ingredients has enabled manufacturers to assure the virtual absence of health risks associated with detergent use. Potential long-term and systemic risks, such as mutagenicity, carcinogenicity, allergy, etc., all are examined and eliminated in the course of the thorough evaluations that routinely accompany raw material considerations during product development. Nevertheless, the possibility remains that finished products could reveal harmful characteristics despite the safety of the raw materials. Acute toxicity and localized reactions are both potential areas of concern, because complex mixtures occasionally show behavior in these respects different from that predicted for the sum of the components.

In fact, the usual commercial detergents display only minimal acute toxicity. Oral ingestion leads usually to vomiting, and this may be accompanied by irritation of the mucous membranes of the gastrointestinal tract and diarrhea. Low toxicity and the fact that vomiting is virtually unpreventable after ingestion of large quantities of detergent essentially exclude any serious threat of poisoning through product misuse, even with respect to children.

Skin irritation is also practically ruled out for modern detergents, provided they are used as directed. Nevertheless, the surfactants that are present and the alkalinity of detergent solutions cause natural lipids to be removed from exposed skin, which can result in minor irritation. Most reported instances of skin irritation can be attributed to extreme levels of exposure coupled with inadequate skin care.

Similarly, no risks appear to be associated with the wearing of articles of apparel that have been treated with modern detergents. Nearly all of the detergent components are removed during rinsing, and the few substances that can be shown to remain on the fabric (e.g., fluorescent whitening agents and fabric softeners) are so tightly bound that only trace amounts can be transferred to the skin, amounts well under the limits of toxicity.

11.3 Conclusions

The constituents of modern detergents can be regarded as toxicologically well characterized and safe for the consumer. The occupational safety concerns that remain in the production process center around surfactants, alkaline components, and the local effects these elements can have on the skin and mucous membranes. Detergent enzymes present the potential for allergenic problems, but the risk is minimal if suitable protective measures are implemented.

From the standpoint of consumer protection, the standard commercial detergents are entirely safe. Even inadvertent misuse of the products can be expected to lead to only minor effects, which are readily reversible.

12 Textiles

A dominant characteristic of the current textile market is the breadth of the available product spectrum. Moreover, essentially all of the fiber types that are worked into textiles for everyday use lend themselves to washing with water — although not all fabrics and finished goods are equally able to withstand high wash temperatures and vigorous mechanical action. Of course, articles that cannot be regarded as washable require dry cleaning; examples include wool suits and dresses, silk neckties, and various highly ornamented items. Nevertheless, the overwhelming majority of fabrics is washable and is washed — in some cases very frequently. Proper washing conditions vary from fabric to fabric, depending on a number of parameters.

Questions regarding washability and, in particular, choice of optimum washing conditions (e.g., selecting the proper cycle with an automatic washing machine), can no longer be resolved simply by determining what type of fibers is involved. Indeed, it has become increasingly necessary to take into account such factors as the types of dyes present and finishing operations to which a fabric may have been subjected.

Dyeing and Printing. All cellulosic fibers (cotton, linen, viscose, etc.) resemble one another in their properties and can all be dyed with the same materials and by similar procedures. What follows is a description of the major types of dyes appropriate to these fibers.

Direct dyes are substances that are transferred onto the fibers directly from a hot dye liquor. Materials that have been subjected to direct dyeing are quite capable of withstanding a gentle wash cycle at 40 °C, although subsequent treatment with special chemicals can lead to a further improvement in both wash- and water-fastness.

If an artificially applied color is to withstand hot water washing (i.e., temperature to 60 °C) and if a direct dye is to be used, the dye must be carefully selected and a special treatment must be administered to increase fastness. Alternatively, dyes can be chosen that lend themselves to after-coppering. In any event, dyes must be chosen from the appropriate fastness grouping at the time of original dyeing.

If a dyed cellulosic fabric is to be subjected to particularly high demands for fastness, a dye of an entirely different type is commonly selected. Among these are vat dyes, reactive dyes, or dyes in the naphthol, phthalocyanine, or sulfur categories. Reactive dyes are capable of forming chemical bonds with cellulose molecules, which display exceptionally high degrees of water fastness.

Only relatively few such dyes are sufficiently versatile for application to wool and silk. Wool is normally treated with acid dyes, at least in the absence of extraordinary requirements for fastness. Metal complex dyes have been specially developed for the treatment of wool, and they possess unusually good fastness properties.

Synthetic and cellulose acetate fibers are generally treated with disperse dyes or, in some cases, pigment dyes. The insoluble coloring agent present is subsequently fixed to the fabric with the aid of heat-setting resins.

The printing of textiles entails the use of essentially the same coloring agents that are employed in dyeing; a separate discussion of this process is, therefore, unnecessary.

Finishing. Consumers have come to expect fabrics emerging from washing with relatively few wrinkles; the wrinkles that remain should disappear quickly without the need for special treatment. Meeting these expectations has required manufacturers to develop special finishing methods involving the use of resins. Appropriate water-soluble resin precondensates are applied to the raw fabric, which are induced to condense during a subsequent drying process.

With the assumption that washing is conducted under the proper conditions, resinfinishing assures that the original crease angle is restored as the fabric dries, thereby eliminating the need for ironing. Furthermore, garments treated in this manner retain their form as they are worn.

Fabric finishes tend not to be permanent: the benefits they confer are gradually lost as a garment is subjected to repeated washing. Certain finishing characteristics can be restored by laundry aftertreatment aids, however, particularly softness, stiffness, or antistatic properties.

In some countries (e.g., the United States and Canada), flameproofing is another form of fabric finishing commonly applied to household goods. Items that have been so treated are generally identified by special labels. Strict adherence to manufacturer's washing instructions is necessary if lack of flammability is to be maintained.

Washable Fabrics. Textile demand for washable fabrics in Europe (i.e., in the countries comprising the EEC) for 1982 is outlined in Table 54 [573]. Three types of material can be seen to account for most of the sales: cotton (40 %), wool (10 %), and synthetic fibers (50 %). The same table also reveals the distribution of these fabrics over the principal categories of household washables. Figure 82 depicts fabric use patterns for the United States for 1976–1984.

In Western Europe, the proportion of white goods in a typical laundry has diminished steadily since the early 1970s, largely as a consequence of change in fashions. This trend is illustrated for the Federal Republic of Germany in Figure 83. Changes in the distribution of fabric types as a function of color, for Germany, are shown in Figure 84. These data can be extrapolated to reflect generally the situation in the European Economic Community as a whole.

By contrast, the distribution pattern between colored and white goods in the United States has been reasonably constant for the last decade (Table 55).

Washing Conditions. Washable woven and knitted fabrics are normally washed either in the household, in a so-called coin laundry (laundromat), or by a profession-

Table 54. Fiber consumption in the domestic production of the most important washable textile articles in the European Economic Community in 1982* (1000 t)

Textiles	Total	Wool	Cotton	Cellulosics	Polyamide	Polyester	Acrylics	Others
Apparel								
Workwear	45.8		32.3	0.8	1.6	10.9		0.2
Skirts	57.7	15.7	13.6	2.4	0.6	20.3	5.0	0.1
Long trousers	53.3	6.1	27.7	2.3	0.9	15.0	1.3	
Sweaters	198.5	38.8	12.3	3.2	5.3	5.7	132.4	0.8
T-shirts and sweater shirts	19.5	0.2	15.0	0.4	0.4	1.4	1.7	0.4
Blouses and shirts	58.0	0.6	26.0	3.8	0.9	24.8	1.9	
Underwear	69.4	2.2	47.3	1.4	10.6	2.5	0.9	4.5
Night wear	34.3	0.2	20.6	1.8	3.5	5.7	2.2	0.3
Footwear	91.9	9.0	10.7	0.6	59.4	0.8	9.5	1.9
Brassieres and foundation garments	6.9		0.9	0.1	4.9	0.2		0.8
Home furnishing								
Towels	59.0		57.7	0.8	0.1	0.4		
Bed linen	138.9		106.2	7.1	4.5	17.9	2.7	0.5
Blankets	52.1	17.9	5.2	4.4	0.6	1.9	18.9	3.2
Net curtains	31.7		0.2	0.2	0.1	28.3	2.9	
Table linen	16.3		11.2	2.1	0.6	1.0	1.4	
	100%	9.7%	41.4%	3.4%	10.1	14.7%	19.4%	1.3%

* Excluding Denmark and Greece.

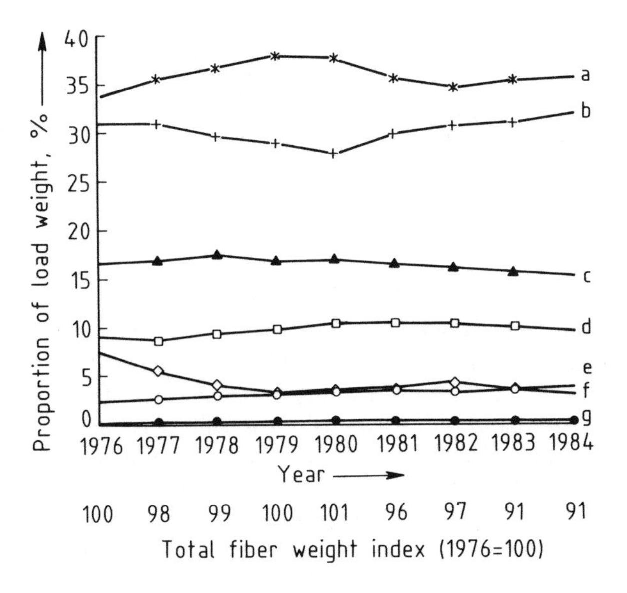

Figure 82. Fiber profile of laundry wash load in the United States
a) Polyester/cotton; b) Cotton; c) Polyester; d) Nylon; e) Rayon and acetate; f) Acrylic and modacrylic; g) Wool

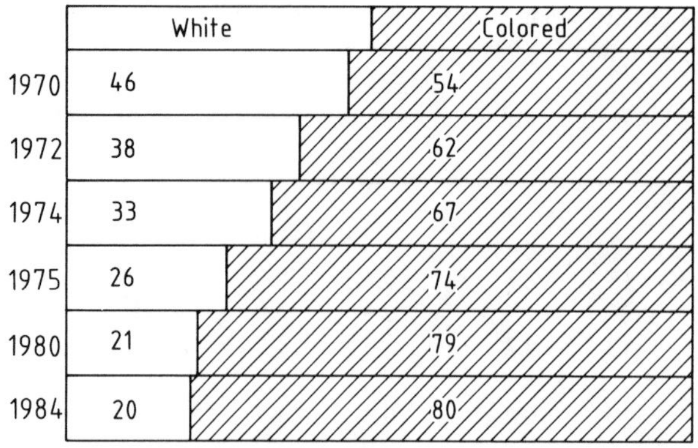

	White	Colored
1970	46	54
1972	38	62
1974	33	67
1975	26	74
1980	21	79
1984	20	80

Figure 83. Proportions of wash loads with respect to colors, % (Federal Republic of Germany) [109]

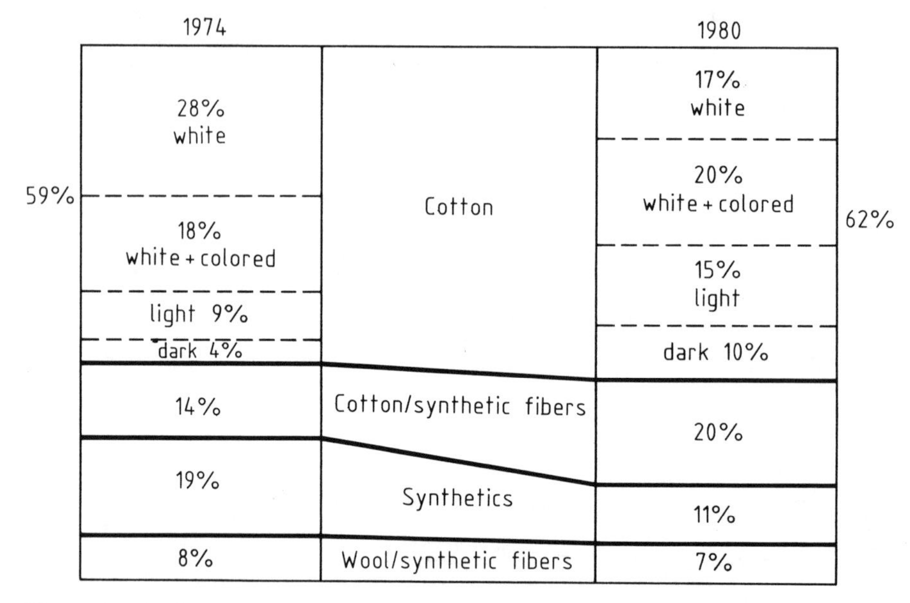

Figure 84. Distribution of fabrics among household washable goods (Federal Republic of Germany) [109]

al laundry [574], [575]. The distribution among the three varies widely from country to country. In the Federal Republic of Germany, for example, more than 90% of household laundry is done at home, usually in a drum-type machine. In Europe generally, wherever the percentage of households with washing machines is high, the fraction of the laundry done at home is also high. A major reason for this pattern is the substantial reduction of effort brought by the technological advances incorporated in current automatic washers. The process has become even simpler with the advent of wash-and-wear fabrics. The finishing work associated with a load of

Table 55. Distribution of colored and white fabrics in the United States

Fabric color (% white)	Percentage		
	1976	1980	1984
All white (100%)	19	15	16
Mostly white (75%)	12	10	13
Mixed white (50%)	20	20	21
Mostly colored (25%)	16	17	16
All colored (0%)	33	37	34
Average % white	42	37	40

laundry currently entails little more than folding and putting away the washed and dried articles (Fig. 85).

Establishing proper washing conditions — temperature, time, mechanical input, wash liquor ratio, and detergent — requires consideration of the characteristics of each of the materials that make up a given article of laundry. For this reason, washability is a characteristic that can only be measured with respect to finished goods. However, conditions imposed during washing, drying, and ironing differ, depending on whether the process is carried out at home or in a commercial laundry. Differences occur in times required for washing and drying, the stress imposed in conjunction with hot-air drying, and the pressure and temperature associated with ironing.

Throughout Western Europe, white and colorfast cottons are usually washed with a heavy-duty detergent and a normal machine cycle, i.e., at ca. 95 °C with a low wash liquor ratio and a standard amount of mechanical input. Cottons that are not colorfast are usually washed at 40–60°C. Only pastels require the use of specialty detergents designed for colored fabrics; their main advantage is the absence of fluorescent whiteners.

Woolens and silks are regarded as especially sensitive to washing conditions. In some cases, these fabrics must simply be accepted as non-washable and submitted to dry cleaning. The surface of a wool fiber has a unique "scaly" construction; this peculiar morphology contributes to wool's characteristic tendency toward *felting*, a term used to describe the appearance of wool after mechanical action has caused individual fibers to become entangled with one another. Wool can be washed much more successfully if it is first given a special treatment known as an anti-felt finish [576]. Woolens and silks should always be washed at low temperature (maximum 30 °C), using a high wash liquor ratio and a reduced level of mechanical input. The detergent employed should be either designed for easy-care fabrics or else a product specifically intended for wool.

Synthetic fabrics based on cellulose, i.e., made from regenerated cellulose fibers (e.g., rayon staple or artificial wool), require a wash temperature of 60 °C, but if sensitive colors are present, the temperature should be reduced to ca. 40 °C. A high

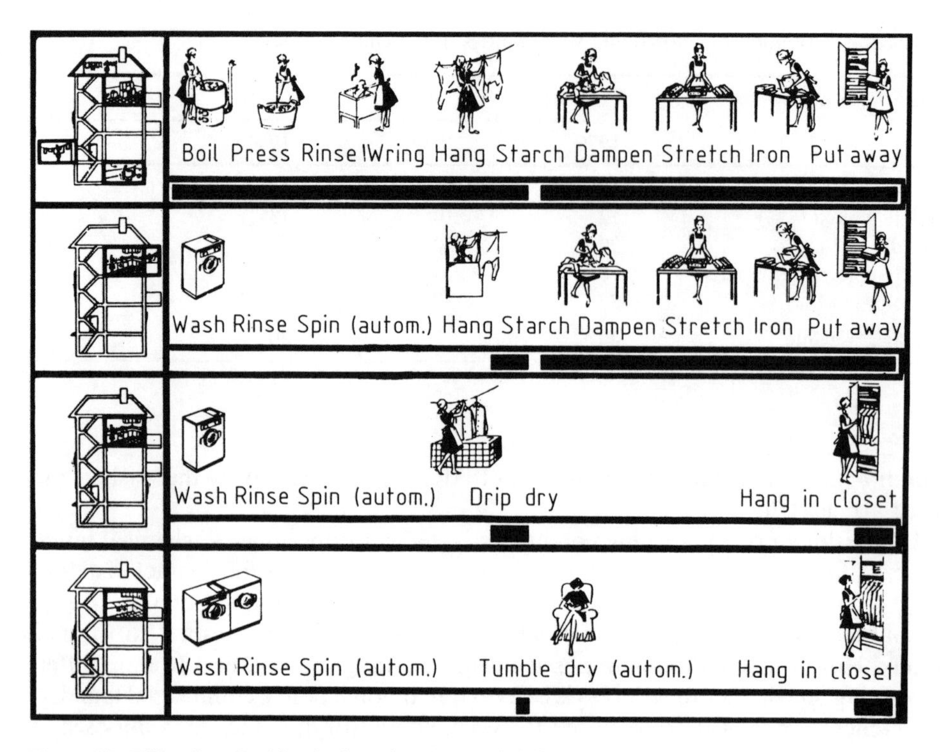

Figure 85. Effort involved in the laundry process [575]
The reduction in energy expenditure accompanying various stages of technological development is symbolized by the length of the black lines under the various pictures. The more automated the process and the more nearly the fabrics show "easy-care" ("wash and wear") characteristics, the less the effort required.

wash liquor ratio is recommended, as is a limited amount of mechanical input. Cellulose acetate is generally washed at 40 °C.

All pure synthetics are washed on a gentle cycle with a high wash liquor ratio. A heavy-duty detergent is considered appropriate, as is one designed for easy-care fabrics. Only in exceptional cases is a special detergent for colored fabrics necessary (e.g., with uniformly colored pastels, when optical brighteners are to be avoided).

White polyamide fabrics can be washed at 60 °C, colored polyamide fabrics at 30 or 40 °C.

White polyester fabrics are washed at 40 or 60 °C, whereas colored polyesters can only withstand a wash temperature of 40 °C.

Knitted fabrics made of polyacrylonitrile are treated like wool; i.e., they are washed in cold water (or at a maximum of 30 °C). White polyurethane undergarments can be washed at 60 °C, and colored items at either 40 or 60 °C, depending on the degree of colorfastness.

Poly(vinyl chloride) fibers have unusually low softening and melting temperatures. Thus, fabrics made of these fibers should never be washed above 30 °C.

Inorganic synthetic fibers (e.g., fiberglass) are used principally in the manufacture of curtains. These require exceptionally high wash liquor ratios. Curtains often are so large that it is impractical to wash them at home in a normal drum-type machine. Dyes employed with fiberglass tend to show poor resistance to scuffs.

Many of the easy-care fabrics on the market consist of mixtures of two or more fiber types brought together as the material is woven or knitted. Such blends are treated like easy-care fabrics, using a high wash liquor ratio and minimal agitation. Recommended wash temperatures vary, depending on the nature and composition of the blend. In general, the recommended laundering conditions are those that apply to the major constituent. Wool/polyacrylonitrile or wool/polyester blends are always washed at low temperature (maximum 30 °C), whereas with resin-finished cotton/polyester blends, 60 °C is appropriate, and even 95 °C is sometimes permissible.

Items that consist of two or more types of fabric or yarn (e.g., linen tablecloths with wool embroidery, cotton shirts with polyamide piping, jackets with an easy-care exterior and a standard rayon lining, etc.) require different treatment than blended fabrics. The same applies to materials containing dyes of varying degrees of fastness (e.g., white polyamide uniforms with brightly colored collars or trim). Such items must always be washed, dried, and ironed under the conditions applicable to the most sensitive fabric present, even if this fabric constitutes only a small portion of

Table 56. Washability of textiles [577]

Fibers	White		Colored		Pastel colored	
	Temperature, °C	Bath ratio *	Temperature, °C	Bath ratio *	Temperature, °C	Bath ratio *
Natural fibers						
Cotton	95	low	40, 60, or 95 **	low	95 or 60	low or high
Linen	95	low	40, 60, or 95 **	low	95 or 60	low or high
Wool	cold, < 30	high	cold, < 30	high	cold, < 30	high
Silk	cold, < 30	high	cold, < 30	high	cold, < 30	high
Chemical fibers (cellulosics)						
Rayon staple	60	high	60 or 40	high	60 or 40	high
Acetate	40	high	40	high	40	high
Chemical fibers (synthetics)						
Polyamide	60	high	30 or 40	high	30 or 40	high
Polyester	30–60	high	30 or 40	high	30 or 40	high
Polyacrylonitrile	cold, < 30	high	cold, < 30	high	cold, < 30	high
Polyurethane	60	high	40–60	high	40–60	high
Poly(vinyl chloride)	30	high	30	high	30	high

* Low: bath ratio 1:5 and normal mechanical input; high: bath ratio 1:20 to 1:30 and decreased mechanical input. ** Colorfast items.

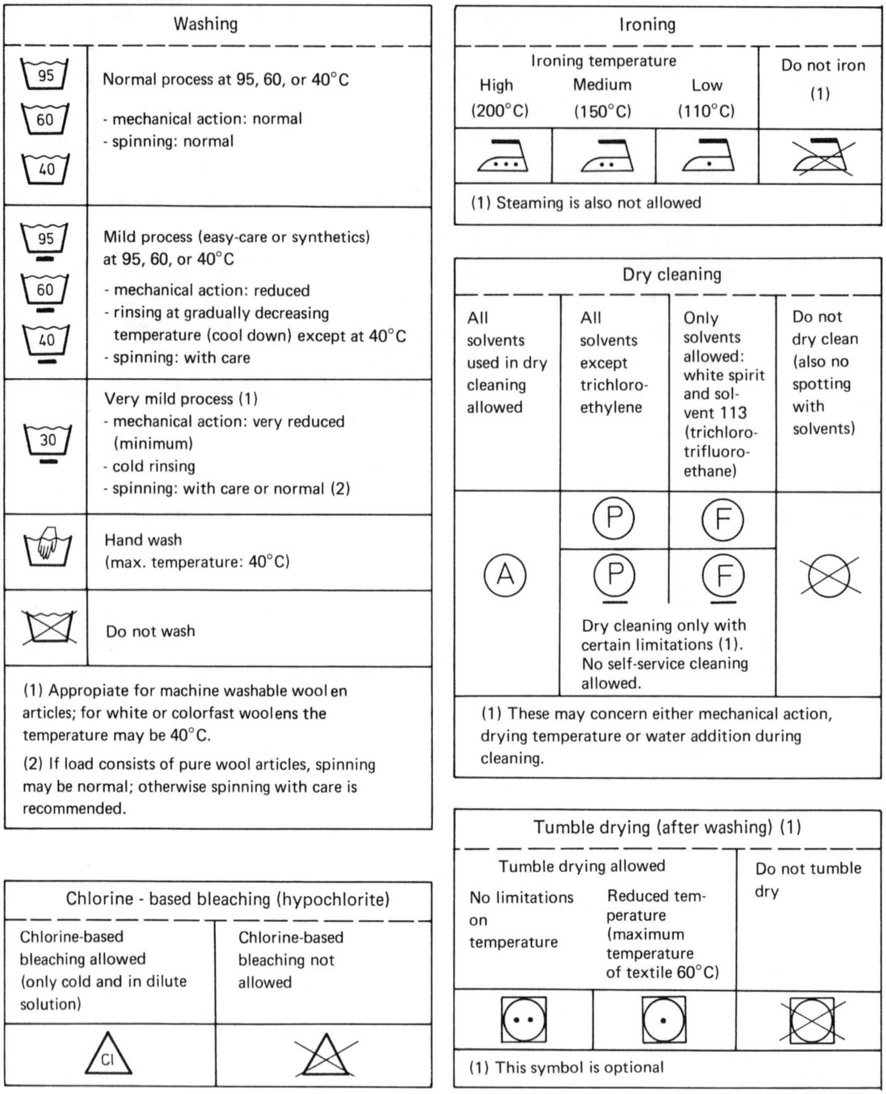

Washing	
95 60 40	Normal process at 95, 60, or 40°C - mechanical action: normal - spinning: normal
95 60 40	Mild process (easy-care or synthetics) at 95, 60, or 40°C - mechanical action: reduced - rinsing at gradually decreasing temperature (cool down) except at 40°C - spinning: with care
30	Very mild process (1) - mechanical action: very reduced (minimum) - cold rinsing - spinning: with care or normal (2)
	Hand wash (max. temperature: 40°C)
	Do not wash

(1) Appropiate for machine washable wool en articles; for white or colorfast woolens the temperature may be 40°C.

(2) If load consists of pure wool articles, spinning may be normal; otherwise spinning with care is recommended.

Chlorine - based bleaching (hypochlorite)	
Chlorine-based bleaching allowed (only cold and in dilute solution)	Chlorine-based bleaching not allowed
Cl	

Ironing			
Ironing temperature			Do not iron (1)
High (200°C)	Medium (150°C)	Low (110°C)	
•••	••	•	

(1) Steaming is also not allowed

Dry cleaning			
All solvents used in dry cleaning allowed	All solvents except trichloro-ethylene	Only solvents allowed: white spirit and sol-vent 113 (trichloro-trifluoro-ethane)	Do not dry clean (also no spotting with solvents)
A	P	F	
	P Dry cleaning only with certain limitations (1). No self-service cleaning allowed.	F	

(1) These may concern either mechanical action, drying temperature or water addition during cleaning.

Tumble drying (after washing) (1)		
Tumble drying allowed		Do not tumble dry
No limitations on temperature	Reduced tem-perature (maximum temperature of textile 60°C)	
••	•	

(1) This symbol is optional

Figure 86. GINETEX international care labeling symbols

the whole. For example, a cotton tablecloth with a small amount of wool embroidery requires the same treatment as if it were made entirely of wool. Table 56 provides a summary of the washability of various fibers.

Care Labeling. The European Economic Community (EEC), acting in the interest of consumers, has developed a uniform and comprehensive system of care labeling for textile goods. This system has required the cooperation of textile fiber manufacturers, the textile and detergent industries, and representatives of the laundry and dry-cleaning trades. The results are illustrated in Figure 86. Labels of the approved type are permanently affixed to items of clothing prior to sale, providing the consumer with a straightforward guide to proper washing and handling. The symbols themselves are internationally protected trademarks, the rights to which are held by the Groupement International d'Etiquetage pour l'Entretien des Textiles (GINETEX). The German trademark rights rest with the Arbeitsgemeinschaft Pflegekennzeichen für Textilien in der Bundesrepublik Deutschland, itself a member of GINETEX. The German organization also confers the right to use these symbols. No license fees are entailed, but the symbols must be used correctly and in their entirety, as specified in the appropriate guidelines. Improper use can lead to loss of the right to employ the symbols. Care labels must be understood merely as sources of advice; they do not in any sense represent evidence of product quality. Nevertheless, any label that appears must be appropriate to the item in question and must be legible and permanently affixed.

Fabric Sorting. In Europe, all laundry is customarily sorted on the basis of the care symbols described above. The situation in the United States is rather different, primarily because of the use there of very different standard washing conditions (cf. Table 19). In particular, wash temperatures in the United States tend to be significantly lower than in Europe due to the absence of separate heating units in American washing machines. Temperature recommendations applicable under these conditions are outlined in Table 57. Sorting of laundry in the United States is usually limited to that required for producing entire loads subject to a common wash temperature, degree of agitation, and spin speed, as well as to the same laundry aid treatment (Table 58).

In Japan, laundry is customarily sorted into three categories, defined solely by the degree of soil: light, normal, and heavy.

Table 57. Wash water temperatures in the United States [578]

Water temperatures	Use
Hot, 130 °F (54 °C) or above	normally soiled permanent-press and synthetic items white and colorfast cottons heavy soils diapers
Warm, 100 °F (38 °C)	dark or noncolorfast colors washable woolens moderate soils knits
Cold, less than 100 °F (38 °C)	lightly soiled fabrics some noncolorfast colors some normally soiled items with extra detergent

Table 58. Cloth sorting in the United States [578]

Color	Amount of soil	Fabric and construction	Lint	
			Donators	Receptors
White	heavy	permanent-press and synthetics	chenille robe, bath towels	corduroy, permanent press, synthetics
Colorfast	normal	towels, jeans, and denims synthetic knits		
Noncolorfast	light	delicates		

Fabric Designation. Current EEC guidelines specify that all textile products offered to consumers must be accompanied by a detailed description of their raw material content. A "textile product" in this regulation is any item whose weight is made up of 80% textile raw materials; i.e., the definition encompasses items that are not totally fabrics in nature.

The guidelines enumerate 39 categories covering the entire range of animal, vegetable, and synthetic fibers, including the newest man-made materials. In this context, the term "synthetic" is not in itself considered definitive; more explicit identification of material category is required (e.g., polyester, polyacrylonitrile, polyamide, polyurethane, etc.).

Raw material content is specified on the basis of a percentage of net fabric weight, e.g.:

70% cotton
20% polyester
10% silk

Constituent textile raw materials whose contribution is less than 10% may be collected under the single heading "other fibers." Thus, a product comprised of 72% cotton, 7% polyester, 7% polyamide, 7% viscose, and 7% acetate would be labeled as follows:

72% cotton
28% other fibers.

13 Washing Machines and Wash Programs (Cycles)

In contrast to the situation in the United States, extensive mechanization of the washing process in Europe was delayed until the late 1950s. A major factor in its ultimate arrival was the development of automatic washers operating on normal house current, particularly those whose spin cycle was sufficiently controlled so that the machine would not need to be permanently fastened to the floor.

Automation in some measure has long been present in institutional laundries. This has culminated in systems designed to operate on a production-line basis, resulting in a multifold increase in hourly output per employee and a major reduction in required personnel. Nevertheless, only a small fraction of household laundry is currently performed commercially, primarily because the effort involved in doing laundry at home has been so dramatically reduced by the advent of the household automatic washer.

13.1 Household Washing Machines

13.1.1 Classification

Two basic types of home automatic washers exist: agitator machines and drum machines. In the United States and in South America and Asia, the agitator-type or the related impeller-type machine is most commonly encountered, and in a form that lacks internal heating facilities and, therefore, must be connected to an external source of hot water if a heated wash liquor is desired. By contrast, this type of machine has lost virtually all of its former significance in Europe, where it has been replaced almost entirely by the drum-type washer (Fig. 87).

Agitator Machines. American agitator-type washing machines have a metal laundry tub, which is usually enamelled. Inside the tub is an open, perforated basket, normally of a size suitable for holding 2–3 kg of laundry.

The laundry tub is filled with 40, 50, or even somewhat more than 60 L of water, depending on the wash cycle chosen and the type and amount of the load. Most manufacturers recommend that detergent be introduced into the machine first, fol-

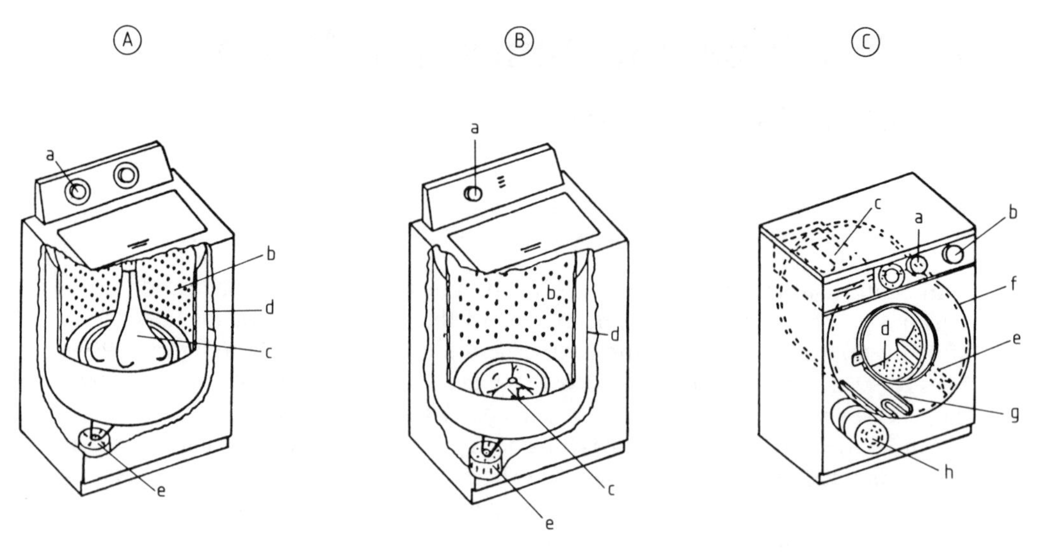

Figure 87. Comparison of U.S., Japanese, and European automatic home washers
A) Agitator washer (United States)
a) Timer; b) Basket; c) Agitator; d) Outer tub; e) Motor
B) Pulsator washer (Japan)
a) Timer; b) Basket; c) Pulsator; d) Outer tub; e) Motor
C) Drum washer (Europe)
a) Timer; b) Thermostat; c) Detergent dispenser; d) Drum; e) Paddles; f) Outer tub; g) Heating coil; h) Motor

lowed by the laundry, which should always be distributed as uniformly as possible. Only then is the selected amount of water allowed to enter. Machines of this type normally have two water connections, one for hot water and one for cold. The desired wash temperature (hot, warm, or cold) is selected by proper setting of the controls, and the selected temperature is achieved by automatic mixing of water from the two input sources. Bleach is not normally included in detergents intended for use in the United States and Japan. Instead, sodium hypochlorite solution is normally added by hand near the end of the wash cycle, although some machines are equipped with automatic dispensers for the purpose.

After the water has been introduced, the agitator begins to function, and the laundry is jostled for ca. 10–15 min by the rhythmic motion of the agitator arms (Fig. 87 A). When the wash cycle is complete, the used wash liquor is pumped off. American machines that are set for the normal cycle then begin a series of spins and spray rinses, whereby fresh cold water is sprayed on the load to remove detergent. Water is again allowed to fill the tub for a brief rinse (ca. 2 min), after which it is once more pumped out and a final spin drying occurs.

Impeller ("Pulsator") Machines. Japanese impeller-type washing machines use a rotating ribbed disk mounted either in the base of the tub or on its side as their source of mechanical input (Fig. 87 B). These machines are invariably smaller than their counterparts in the United States, with a capacity limited to 1–1.5 kg of laundry.

Drum Machines. In automatic drum-type washers, the laundry is placed in a horizontal, perforated drum, which rotates on its axis and in alternating directions (Fig. 87 C). For a normal cotton wash, only the lower third of the drum is filled with wash liquor (ratio ca. 1:5), which means that, in contrast to agitator machines, the laundry is not fully submerged. Individual laundry items are repeatedly lifted by paddles located on the edge of the drum, each time falling again into the wash liquor for renewed soaking, rubbing, and compacting [579], [580]. The wash liquor is heated internally by means of electrical heating coils located at the bottom of the bath. Some machines are equipped with a separate hot water inlet for use in households that have access to an economical external hot water source. In this case the bath is often tempered initially to 30–40 °C by introduction of cold water, thereby preventing proteinaceous soil from being thermally set in the fabric. Subsequent heating can then take place if this is required by the particular automatic cycle selected. Laundry is loaded either through a door in the side of the drum or through its open front. Machines of these two types are called top loaders and front loaders, respectively.

Detergent is introduced through a dispenser. In a front loader, the dispenser is usually a device resembling a drawer located at the front of the machine. The dispenser for a top loader generally consists of a series of compartments located beneath a door in the top of the machine. Such dispensers permit automatic sequences consisting of either one or two wash cycles (i.e., a prewash and a main wash). Additional compartments are also provided for the automatic introduction of laundry aftertreatment aids or, especially in South Europe, automatic dosage during the rinse cycle of a chlorine-containing bleach.

Semiautomatic drum washers are machines that perform only the wash and rinse steps, requiring purchase of a separate spin dryer. So-called twin tub washers retain

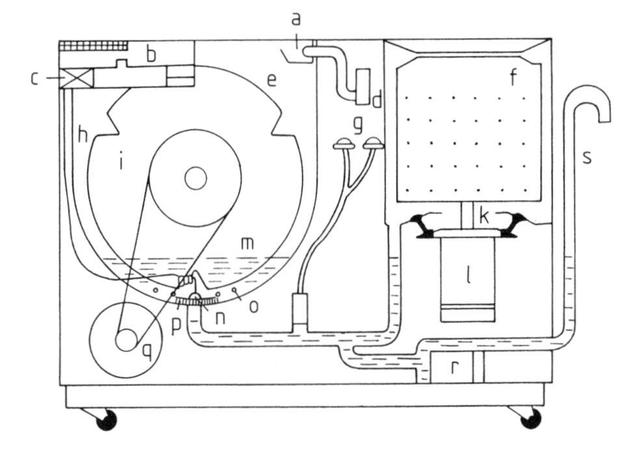

Figure 88. Twin tub automatic washer
a) Detergent dispenser; b) Operating controls; c) Thermostat; d) Magnetic water inlet valve; e) Drum closure; f) Spin basket; g) Membrane switches; h) Laundry tub; i) Washing drum; k) Spinner suspension; l) Spinner motor; m) Wash liquor; n) Thermostat sensor; o) Heating coils; p) Lint filter; q) Drive motor; r) Electrical suds pump; s) Drain hose

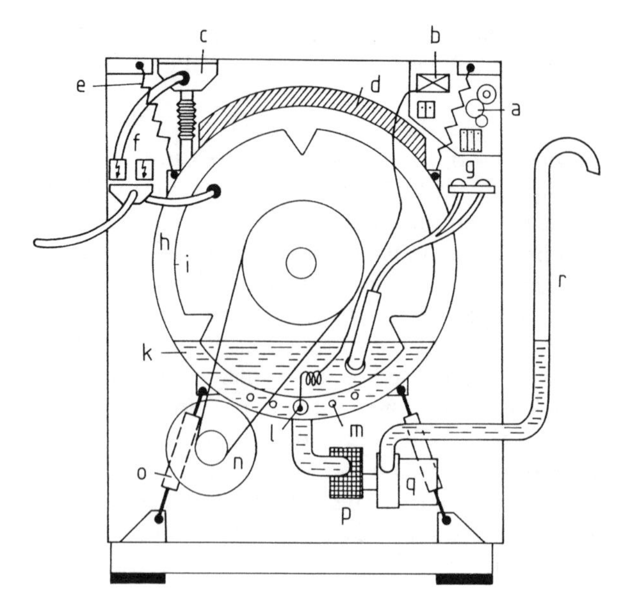

Figure 89. Fully automatic drum-type washer
a) Control panel;
b) Thermostat; c) Detergent dispenser; d) Counterweight;
e) Spring suspension;
f) Solenoid water inlet valve;
g) Membrane switches;
h) Laundry tub; i) Washing drum; k) Wash liquor;
l) Thermostat sensor;
m) Heating coils; n) Drive motor (for both wash and spin cycles); o) Shock absorbers;
p) Lint filter; q) Electrical suds pump; r) Drain hose

the concept of the two separate functions of washing and spin drying, but combine the appropriate devices in a single cabinet (Fig. 88). Such machines provide particularly effective spin drying, since the spin drum is mounted vertically and operates at a higher rotational rate than is possible with a dual-function horizontal drum.

Fully automatic drum washers utilize the same drum for successively accomplishing three separate operations: washing, rinsing, and spin drying. Imbalance problems that occur during the spin cycle required that machines built before ca. 1962 be firmly anchored to the floor, although free-standing machines are now common. The newer models are equipped with special mechanical devices that compensate for load imbalances, thereby obviating the need for permanent installation (Fig. 89). The trend toward space-saving machines has led to the development of smaller washers and fitted washers for built-in kitchens.

13.1.2 Operational Parameters [581]

Four factors must be regulated in the washing operation: washing chemistry, mechanical input, wash temperature, and time.

The effect of each factor on wash performance varies, depending on the washing techniques employed. Figure 90 illustrates graphically by a Sinner's circle the percentage influence of the individual parameters on the overall process.

The inner circle implies that the effectiveness of the four factors is a consequence of the medium that unites them, i.e., water.

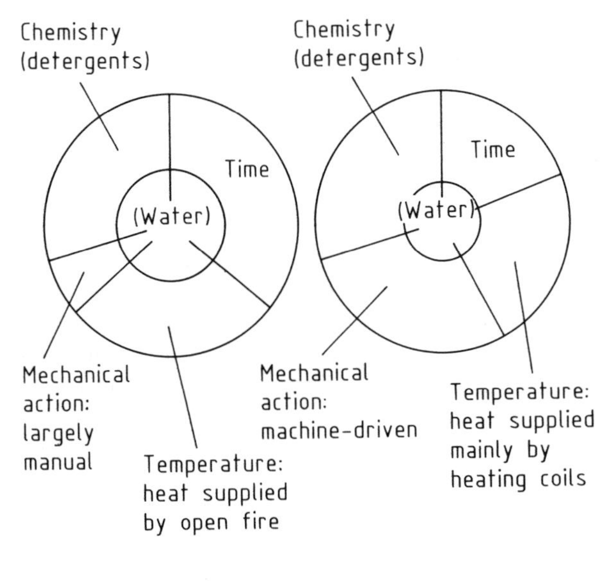

Figure 90. Circular laundry chart (Sinner's circle) [581]

In the days when laundry was done in an open vat, the required amount of water was large, the role of mechanical action was very limited, and time was a very important factor. A washing machine provides considerably more mechanical action; thus, the wash time can be shortened.

Wash Liquor Ratio. "Wash liquor ratio" (bath ratio) means the ratio of dry laundry (in kilograms) to the volume of wash liquor (in liters). The total amount of wash liquor required in the overall process is made up of two portions: that which is absorbed by the laundry (bound wash liquor) and that which remains in excess (free wash liquor).

Various washing processes have different wash liquor requirements. Vat washing generally involved ca. 10 L of water for 1 kg of laundry, a wash liquor ratio of 1:10. An agitator-type washing machine typically uses 15 L of water per kg of laundry, a wash liquor ratio of 1:15, whereas an impeller machine requires a ratio of ca. 1:20. A drum-type automatic washer is characterized by a very low 1:5 wash liquor ratio with cotton, but rather high values of 1:20 or 1:30 for easy-care fabrics.

Wash Liquor Level and Reversing Rhythm. A drum-type washer is programmed to rotate in a reversing fashion, first in one direction and then in the other. The extent of the mechanical action imparted to the laundry can be altered by changing either the rhythm of alternation or the wash liquor level (or both). A low wash liquor level (wash liquor ratio of ca. 1:5) and relatively rapid reversal (e.g., a reversing rhythm of 12 s of rotation/4 s of pause) causes a large mechanical effect on the laundry. In contrast, a high wash liquor level (wash liquor ratio 1:20 to 1:30) and slower reversal (reversing rhythm 4 s of rotation/12 s of pause) provides less mechanical action, hence the term "gentle cycle."

Residual Moisture. Residual moisture is defined as the amount of water remaining in the laundry after draining. Data are reported as a percentage of the mass of air-dried laundry. Separate spin drying with a vertical drum permits relatively high rates of rotation: up to ca. 2800 revolutions/min. Corresponding residual moisture levels are typically ca. 40–50%. Automatic drum-type machines spin dry the load at rates of 600–1200 revolutions/min and result in residual moisture levels of 100–70%.

Wash Temperature. In Western Europe clothes are commonly washed over the wide temperature range of 30–95 °C, whereas elsewhere much lower temperatures are used (e.g., in the United States to ca. 55 °C, in Japan only to 25 °C or, in exceptional cases, 40 °C). The high wash temperatures used in Europe are based largely on tradition and on the firmly held belief that "only clothes that have been boiled are really clean." Nonetheless, this attitude is changing as a result of the increasing popularity of colored and permanent-press fabrics, the desire to conserve energy, and the introduction of effective heavy-duty detergents, designed for use at a lower temperature (cf. Section 4.1). Lower temperatures have long been the rule in the United States and in Japan, a fact partly explained by the hygienic function associated with use of chlorine-containing bleaches.

Wash Times. Customary wash times also differ substantially between Europe on the one hand and the United States and Japan on the other. Much of the difference is attributable to differences in washing technique: relatively low wash temperatures and the absence of built-in heating units in the United States and in Japan, as compared to Europe, where a typical washer can heat water to 95 °C. The heating process also consumes a great deal of time. Nearly an hour is required to raise the temperature of the wash liquor in a drum-type machine to a preset value of 95 °C, not counting the time occupied by a prewash. A wash cycle that includes a prewash entails discarding the first wash liquor after it reaches ca. 45 °C, which adds ca. 15 min to the process. When the time required for the rinse and spin dry operations is also taken into account, a normal 95 °C wash operation can consume a total of ca. 2 h.

The cycles provided for most drum-type automatic washers use temperature as their controlling variable. Thus, a high wash temperature dictates a long wash cycle whereas a low temperature requires less time. Recently it has become common for manufacturers to install optional "energy-saving cycles," where a lower wash temperature is compensated by increasing the wash time. The resulting increase in time and mechanical input accomplishes essentially the same end as a high-temperature wash; the corresponding Sinner's circle reveals that a temperature deficit has been balanced by a gain in time and mechanical action.

Washing machines in the United States and in Japan operate considerably more rapidly. These normally require only 10–15 min for the actual wash portion of the overall laundry cycle.

13.1.3 Wash Cycles

Since laundry habits and washing machines vary considerably in different parts of the world, discussion of washing cycles under separate headings reflects the three major traditions: those found in Japan, the United States, and Western Europe.

13.1.3.1 Japanese Washing Machines and Laundry Habits

Approximately 75% of the washing machines found in Japanese households are of the impeller type, the remaining 25% being equipped with agitators. About two-thirds of the machines are of the twin-tub variety, in which a single housing contains separate tubs for washing and spin drying, with the impeller or agitator located in the former. A characteristic of all Japanese washing machines is the absence of any provision for temperature control, since the process is always conducted at the inlet temperature of the tap water. The set of wash cycles programmed into a Japanese machine also provides for the Japanese custom of repeated use of a single batch of wash liquor.

The most important variables with a Japanese washing machine are the water level and the extent of mechanical action. Washing time is adjusted to match the degree of soil of the laundry (ca. 25–35 min). Machines other than the twin-tub variety offer a choice of the following three basic termination procedures (cf. Tables 19 and 59):

drain and spin
drain, rinse, and spin
no spin, drip dry (i.e., the machine stops at the end of the cycle without pumping out the water)

13.1.3.2 American Washing Machines and Laundry Habits

Although some drum-type machines exist in the United States, the automatic agitator washing machine is by far the most common in American households. Such machines carry out a full set of laundry operations in the usual sequence, i.e., wash, rinse, and spin dry. American machines resemble their Japanese counterparts in that they lack heating coils; however, American machines are designed for connection to

Table 59. Program selection of a typical Japanese automatic washer

Program	Application*
Program selector	
Automatic program	
Gentle	for lightly soiled clothes (15 min)
Normal	for normally soiled clothes (25 min)
Heavy	for heavily soiled clothes (29 min)
Optional program	
Wash only	for reuse of wash water and detergent if the amount of clothes is large enough to separate into more than one wash load (9 min)
Drain–rinse–spin	for initiating a rinse of clothes after they have been washed (21 min)
Drain–spin	to be used when only spinning is desired (6 min)
Wash selector	
Gentle	for delicate fabrics such as wool, silk, and delicate synthetics
Regular	for normal synthetics, mixed fibers, and sturdy fabrics such as cotton and linen
Spin selector	
Full cycle	the machine automatically performs all normal washing operations: wash, drain, and spin
No spin, drip dry	the machine ceases operation without draining the tub and without spinning; delicate fabrics can be removed to prevent damage which may be caused by spinning; to activate spinning, set the spin selector to "full cycle" and press the "drain–spin" button; to reuse water left in the tub, press any desired program button

* In parentheses: total time of program.

Table 60. Typical wash cycles in the United States

Wash cycles	Fabric	Soil	Wash time, min	Water temperature	
				Wash	Rinse
Regular heavy	sturdy	heavy	10–14	warm or hot	cold
		moderate	8–14	warm	cold
		light	4–14	cold or warm	cold
Knits gentle	synthetic delicate	light	4–8	warm	cold
		moderate	4–6	warm	cold
Permanent press	regular	heavy	8–10	warm or hot	cold
		light to moderate	4–10	warm	cold
Soak	all colorfast	heavy	22	cold or warm	none
		stained	22	cold or warm	none
Prewash	all colorfast	heavy	6	cold or warm	none
		stained	6	cold or warm	none

domestic sources of both hot and cold water. Three wash temperatures are normally provided: hot, warm, and cold. Rinse temperatures can also be selected, usually from among the two alternatives, warm and cold. One further variable is the wash liquor level, which is adjusted according to the amount and type of laundry to be washed. The mechanical input is also adjustable within limits by introducing changes in the agitator speed and the wash time. The following set of cycles is typical (cf. Tables 19 and 60):

regular (heavy-duty wash)
permanent press
knit or gentle
soak
prewash

13.1.3.3 Western European Washing Machines and Laundry Habits

Drum-type washing machines in which the water can be heated to 90–95 °C dominate the household market in Western Europe. To meet the demands posed by current fabrics, such machines are provided with a wide variety of wash cycles. In choosing the appropriate cycle for a given load of laundry, the first decision is whether or not a prewash is desired. Options in later stages are supported by the presence of a special third compartment in the detergent dispenser, which permits addition of aftertreatment aids to the final rinse. Machines built for the Mediterranean market commonly offer a fourth compartment, which allows addition of chlorine bleach to the second rinse.

Selection of the proper cycle is accomplished in one of several ways. Some machines are equipped with one control knob for establishing the cycle and a separate stepless temperature regulation control. Alternatively, both functions can be combined in a simplified "one button" system. Other models offer a set of pushbutton controls, while some machines even incorporate a system in which the user engages in a kind of "dialogue" with the control panel.

Separate temperature control provides the greatest degree of flexibility, since even the dialogue machines place limits on the available temperature settings once certain fundamental parameters have been established.

Dialogue automatic washers currently represent the latest state-of-the-art with respect to developments in this sector. The introduction of dialogue systems was made possible by the availability of miniaturized electronic components, especially microprocessor chips, which in turn control all machine functions. Operation of such a machine requires the user to engage in a brief "conversation" with the machine by responding to various instructions or inquiries posed by an LED display panel. The questions are centered around the amount and type of clothes to be washed, their degree of soil, and the user's preferences with respect to various options that can be incorporated into the laundering process. Thus, the cycle that results is tailored to the specific situation at hand, and all the related parameters are appropriately monitored by the electronic control unit to minimize the consumption of water and electricity.

Management of any wash cycle requires a more or less complicated switching system. Very few machines (mostly those in the lowest price category) continue to rely on time-based switching, an approach in which each operation is allowed to continue for a precise and predetermined length of time. The principal disadvantage of such a system is that temperature cannot be properly established without taking into account the temperature and volume of the water being introduced and the temperature of the surroundings, all of which can vary.

In most modern European washing machines, the control system is based on the temperature of the wash liquor, thereby assuring that the desired temperature is reached at each stage of the operation. Complete adaptation is thus assured to whatever water supply is available. Thus, the length of time involved in heating the water becomes variable, as does the overall length of a cycle.

The very latest equipment is characterized by the presence of integrated electronic circuitry based on microprocessors, some of which are so complex that they might be regarded as microcomputers. These circuits assume full functional control of the machine. So-called fully electronic washing machines of this type lack any complex mechanical switching formerly typical of an automatic washer.

Garments and fabrics sold in Europe contain uniform labels specifying the use of one of three (or four) wash temperatures: 30 °C (or 40 °C), 60 °C, and 95 °C. Table 61 gives examples of the wash cycles commonly provided to meet these needs (cf. also Table 19).

13.1.4 Energy Consumption

Mechanization of the washing process has relieved much of the burden formerly resting on the person doing the wash. At the same time, however, steady increases in energy costs have made use of a household drum-type washing machine increasingly expensive. In particular, the cost increases as the selected wash temperature is increased (Table 62).

It is not surprising that efforts to reduce the costs have begun with the 95 °C cycles, since these hold the greatest potential for energy savings (e.g., by lowering the water temperature from 95 °C to 60 °C). Some washers have incorporated an optional energy conservation cycle. Such cycles are usually designed to function at 60 °C. Examples of typical energy conservation measures are shown in Table 63. (For information on the related trend in the United States toward washing at lower temperatures, see Fig. 59 in Section 4.1.)

Table 61. Typical wash cycles for a European drum-type automatic washer [581]

Cycle designation	Application	Prewash temperature, °C	Main wash temperature, °C	Drum operation		Water level		Number of rinses	Final rinse water retained	Spin drying	
				Normal	Gentle	Low (16–20 L)	High (25–30 L)			Spin before removal	Normal
Normal ("boiling") cycle, 95 °C	heat-resistant items (cotton, linen, and blends)	45	90–100	+		+		4–5			+
Easy-care, 95 °C	heat-resistant cottons, permanent press	45	90–100		+	+		3	+	+	
Colored fabrics, 60 °C	nonfast colors (cotton, linen, and blends)	45	60	+		+		4–5			+
White easy-care, 60 °C	white synthetics (including polyamides, polyesters, and cotton blends)	45	60		+		+	3	+	+	
Colored fabrics, 30 °C	sensitive colors (cotton, linen, and blends)	30	30	+		+		4–5			+
Colored easy-care or delicate washables, 30 °C	colored synthetics, e.g., polyamides and polyesters; polyacrylonitrile, polyacrylonitrile/wool, rayon, acetate, and cupro	30	30		+		+	3	+	+	
Wool, silk, 30 °C	washable woolens		30		+		+	2–3	+		
Fabric softener	normal and colored fabrics			+		+					+

Table 62. Typical energy consumption for various wash cycles with a European drum-type automatic washer [582]

Program	Energy consumption, kWh
95 °C Boiling wash program	3.2
60 °C Colored fabrics	1.8
60 °C Synthetics	1.6
40 °C Synthetics	0.8
30 °C Woolens	0.4

Table 63. Examples of energy-conserving cycles offered with European automatic washers [582]

Modification of program	Degree of laundry soiling	Wash load, kg	Savings in comparison to a normal boiling wash program, %	
			Electricity	Water
No prewash, one rinse cycle less	lightly soiled laundry	4	11–15	11
Reduction of the washing temperature from 95 °C to 60 °C	lightly soiled laundry	4	35–50	
Reduction of the washing temperature from 95 °C to 60 °C, with extension of the 60 °C washing time by ca. 30 min	lightly soiled laundry	4	30–45	
95 °C or 60 °C program with reduced water level	moderately soiled laundry	2	20–45	25–40

13.1.5 Construction Materials Used in Washing Machines

Materials used in the construction of washing machines must exhibit a high degree of resistance to various mechanical and chemical influences. The following materials are particularly well-suited to the purpose:

Stainless Steel. Virtually all inner drums are currently made of stainless steel. Chromium steel continues to be widely used for the construction of wash liquor containers, dispensers, lint filters, heating coils, and thermostat sensors.

Sheet Steel. Cabinets, base plates, and tops are fabricated from sheet steel, although a protective surface must be provided (e.g., enamel paint, baked enamel, galvanization). Baked enamel provides the most permanent form of protection, and sheet steel with a baked enamel finish is often regarded as suitable for use even in the wash liquor compartment.

Plastics. Certain components can be made from alkali-resistant plastics, such as polypropylene. Typical applications include detergent dispensers, paddles, operating controls, mechanical parts, and various small components. Recently, glass fiber reinforced polypropylene has also been employed in drum-type machines for the wash liquor compartment.

Rubber. Molded rubber is the material of choice for components that require a high degree of elasticity. This applies particularly to gaskets, such as those around the door, as well as to various hoses, hose connectors, and shaft seals. Synthetic rubber is used exclusively for such purposes. One disadvantage of rubber is its tendency to deteriorate with age. Deterioration is accelerated by exposure to grease, by which the rubber gradually acquires a soft and greasy character.

13.1.6 The Market for Washing Machines

The percentage of households equipped with washing machines varies considerably among the countries of Western Europe. Percentage estimates for 1982/1983 are as follows:

Austria	81
Belgium	86
Federal Republic of Germany	90
France	82
Great Britain	88
Italy	87
Sweden	65
Switzerland	50
The Netherlands	87

In 1983, 1.6×10^6 automatic washers were produced in the Federal Republic of Germany alone. Of these, 679 000 were exported, but an additional 488 000 were imported [583]. Front loaders accounted for 80% of the total, with the remaining 20% being top loaders [584].

Worldwide, the United States represents the largest washing machine market. Nevertheless, market penetration in the United States is lower than that in several European countries and Japan. Only ca. 74% of American households had their own washing machine in 1984 [585]. In the same year, the United States was responsible for the manufacture of 5.05×10^6 units [586].

Virtually every household in Japan is equipped with a washing machine [585]. The Japanese manufactured 4.72×10^6 units in 1982. Of these, 922 000 were fully automatic, whereas 3.8×10^6 were simpler types. Japanese washing machine export in 1982 was 1.3×10^6 units, most of which were destined for neighboring countries in Asia [587].

13.2 Laundry Dryers

Laundry dryers represent a further stage in the development of laundry technology. Dryers are machines that utilize an influx of heat to cause evaporation of the moisture remaining in spin-dried laundry. Electricity is normally used to supply the heat, although some dryers in the United States are heated by gas. Several basic types of dryers exist, including those based on circulation of heated air, heat transfer from metal surfaces, and radiant heating. Most are drum-type machines, so-called tumblers, and they can be categorized by the way they dispose of the resulting steam:

Ventilated Dryers. Dryers in this category function on the same principle as a fan. Room air is drawn in by suction, passed through a heated zone, and then introduced into the laundry. As the air becomes enriched in moisture, it is vented through a lint filter either directly to the room or out of the building through an exhaust duct.

Condensing Dryers. *Condensing Agent: Room Air.* In such dryers, the air needed for drying is drawn from the room and passed over a heating device before encountering the laundry. However, the moist air leaving the laundry chamber is cooled by an additional supply of incoming room air. This cooling takes place in the presence of a drip screen, which collects precipitated moisture. The air returns to the room at a temperature of ca. 50 °C. Water from the drip screen collects in a condensate vessel and must be emptied after each drying cycle.

Condensing Agent: Fresh Water. Dryers of this type require not only a supply of electricity, but also connections to a water line and a drain. The moist, hot air emerging from the laundry is either sprayed directly with cooling water or it is passed through a cascade of cooling water droplets. Condensed moisture is removed by a pumping system.

Combination Washer–Dryers. The European washer–dryer is a further development of the drum-type washing machine. It is comprised of a washer combined with a water-cooled condensation dryer, and the same drum is used for washing, spinning, and drying. One disadvantage of these machines is the fact that the small size of the washing drum requires a full load of laundry to be divided into two batches before it can be dried. The heat necessary for drying is supplied either by radiation or by direct contact.

Market Outlook for Laundry Dryers. Market penetration for laundry dryers in Europe in 1983 was only ca. 10%. The Federal Republic of Germany is typical; by late 1983 the level of dryer market penetration was 11%, scarcely greater than the European average. By contrast, the corresponding figure for Great Britain was ca. 25%, with other leaders in this respect being Belgium, Switzerland, The Netherlands, and most of the Scandinavian countries, leaving the Federal Republic of Germany in 1983 in ninth place within Europe [588]. In France, Italy, and other countries in the south of Europe, market penetration is generally 1% or less. The Federal Republic of Germany produced 380 000 dryers in 1983, of which 145 000 were exported; imports in 1983 totaled 90 000 units [583]. With respect to type, 68% were ventilated machines and the remaining 32% employed condensation technology [588].

The situation in North America is quite different from that in Europe. In the United States, 66% of all households are equipped with electric dryers, as are 60% of the households in Canada [588]. United States production of household electric dryers in 1984 was 2.9×10^6 units, in addition to which ca. 750 000 gas-heated units were produced [586].

Although nearly all Japanese households are estimated to have washers, the fraction with dryers is only ca. 1%. Nevertheless, the trend is clearly toward increasing dryer sales. Many Japanese consumers are becoming convinced that dryers are a worthwhile investment, partly because of the high population density in Japanese metropolitan areas. However, another factor is the Japanese climate, with its traditionally rainy spring, during which outdoor drying is difficult.

13.3 Washing Machines for Institutional Use

[160], [589], [590]

Washing machines designed for use by commercial laundries are generally designed to keep operating costs to a minimum. As a result, the more labor-intensive batch laundry systems are being increasingly replaced by more efficient continuous devices. In contrast to European households, where laundry water is heated electrically, commercial laundries worldwide usually employ high-pressure steam as their source of heat. Heat from the steam is led into the wash liquor either directly or indirectly, resulting in more rapid heating and shorter wash times.

13.3.1 Batch-Type Machines

There are many situations even today in which batch-type commercial washing machines are indispensable. This is particularly true where a frequent need exists to process numerous small batches of different kinds of laundry. Batch-type machines require a substantial amount of manual supervision; thus, their use is expensive.

Conventional Drum-Type Devices. Such devices are basically scaled-up versions of household washing machines. Commercial front loaders often allow the processing

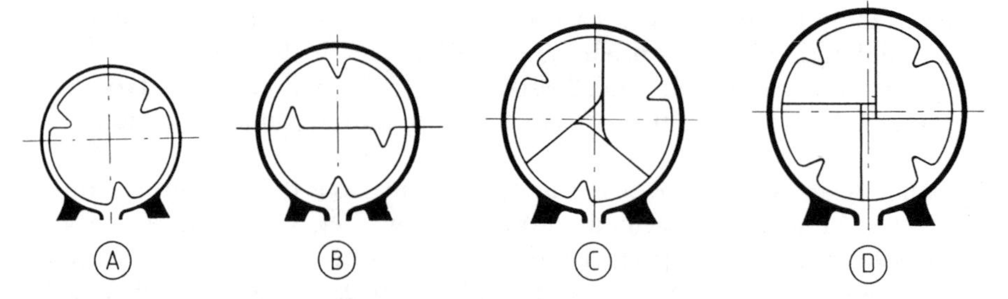

Figure 91. Four common designs for commercial washing machines [590]
A) Open machine; internal drum diameter up to 900 mm; inner drum lacking subdivisions;
B) Pullmann type machine; internal drum diameter 900–1400 mm; inner drum divided into two equivalent compartments by means of an axial partition;
C) Y-type machine; internal drum diameter 1400–1700 mm; inner drum divided into three compartments by three axial partitions joined along the axis;
D) Star-pattern machine, internal drum diameter >1700 mm; inner drum divided into four compartments by four axial partitions joined along the axis

of only up to 50 kg of laundry, whereas top loaders may handle up to 600 kg. Reversal and rotation rates commonly produce tangential velocities of 1.0–1.5 m/s. The large drums require that the effect of mechanical action resulting from great fall heights and excessive weight be moderated and that loading and removing the laundry be simplified. These targets are generally accomplished by introducing into the drum a set of partitions that divide the drum into compartments. Three partitioning schemes (the Pullmann, Y-, and star-pattern systems) are illustrated in Figure 91.

Machines of this type are controlled by program cards. Water is removed from the compartments at the end of the wash and rinse cycles, after which the laundry is transferred to large spin dryers. Final drying is also performed in large commercial-type equipment.

Wash Extractors. Machines in this category, with a capacity of up to 360 kg, are capable of not only the wash step, but also a spin drying and "fluffing" of the laundered items. Spinning occurs after each wash and rinse operation, an expedient that has the effect of minimizing the number of required rinses. This in turn conserves water and reduces overall wash time to ca. 40 min (Fig. 92). Wash extractors are used in all parts of the world, although the principal manufacturers are firms in the United States.

13.3.2 Continuous Batch Washers

The development of the continuous batch washer is a consequence of the demand for a laundry system requiring even less manual intervention. Such machines operate

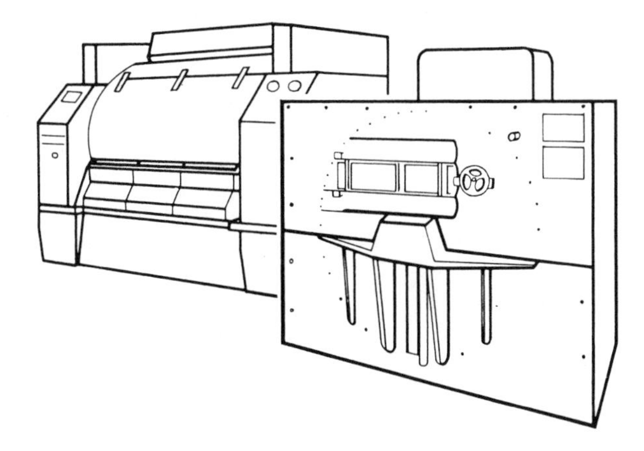

Figure 92. Schematic diagram of a wash extractor [590]

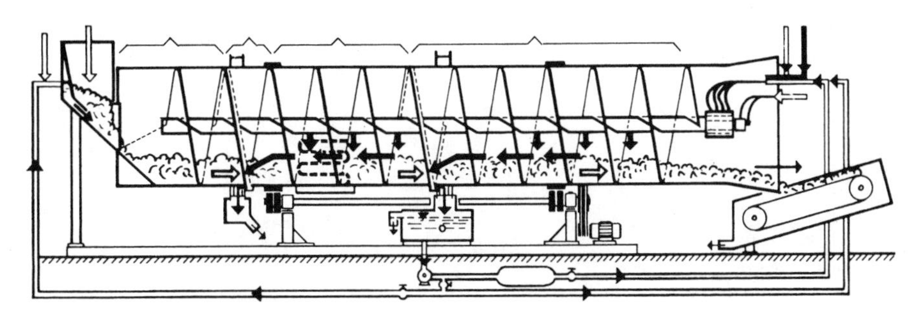

Figure 93. Voss–Archimedia continuous batch washer [590]

uninterrupted and are best suited to institutions faced with large amounts of relatively uniform laundry. Soiled laundry is loaded onto either a conveyor belt or a suspended conveyor system. As it progresses through the machine, the laundry is exposed to a series of zones designed to accomplish the equivalent of wetting, prewashing, main washing, and rinsing. Each zone operates on the countercurrent principle: all water encountered by the laundry flows continuously and in a direction opposite to the motion of the laundry. Each zone leads directly into the next, resulting in a significant saving in energy, water, and detergent.

Following the actual washing process, the laundry is transported by means of a ramp or a conveyor belt to a spin dryer, a water extraction press, or a set of wringers, where excess water is removed; the damp goods are then passed to some type of drying device. The entire process occurs without manual intervention. Continuous batch washers are especially widespread in Central Europe, although recently they have increased in popularity in Japan and in the United States. Figure 93 illustrates the design of a typical continuous batch washer, in this case one of the Voss–Archimedia type.

14 References

General References

K. Lindner: *Tenside-Textilhilfsmittel-Waschmittel-Waschrohstoffe,* Wissenschaftl. Verlagsges. Stuttgart, vols. I and II, 1964, vol. III, 1971.

A. M. Schwartz, J. W. Perry, J. Berch: *Surface active Agents,* vols. I and II, Interscience Publ. Inc., New York 1958.

Kirk-Othmer, **22,** 332–432.

J. Falbe: *Surfactants in Consumer Products,* Springer Verlag, Heidelberg 1987.

Kao Corporation: *Surfactants,* Chuo Toshado Printing Co., Tokyo 1983.

H. Bueren, H. Grossmann: *Grenzflächenaktive Substanzen,* Verlag Chemie, Weinheim 1971.

G. Gawalek: *Tenside,* Akademie-Verlag, Berlin 1975.

M. J. Rosen: *Surfactants and Interfacial Phenomena,* J. Wiley & Sons, New York–Chichester–Brisbane–Toronto 1978.

H. Stache: *Tensid-Taschenbuch,* 2nd ed., Hanser Verlag München–Wien 1981.

E. H. Lucassen-Reynders: *Anionic Surfactants,* Marcel Dekker, New York 1981.

K. Laux et al.: „Tenside," in *Winnacker-Küchler,* 4th ed., vol. 7 (1986) pp. 84–148.

P. Berth, J. Heidrich, G. Jakobi: „Der Einsatz von Tensiden in Waschmitteln Gestern-Heute-Morgen," *Tenside Deterg.* **17** (1980) 228–235.

B. Werdelmann in: *Proceedings of the World Surfactants Congress München,* vol. I, Kürle Verlag, Gelnhausen 1984, p. 3.

H. Andree, P. Krings, H. Verbeek: „Tenside und ihre Kombinationen in zukünftigen Waschmittelformulierungen," *Seifen Öle Fette Wachse* **108** (1982) 277–282.

K. J. Bock, H. Stache: "Surfactants" in: *The Handbook of Environmental Chemistry, 3,* Part B, Springer Verlag, Berlin–Heidelberg–New York 1982, pp. 163–199.

Henkel KGaA: *Fettalkohole, Rohstoffe, Verfahren und Verwendung,* Düsseldorf 1981.

H. R. Jaag: *Über proteolytische Enzyme, deren Prüfmethoden und Einsatzmöglichkeiten in der Waschmittelindustrie,* Verlag Lang, Bern 1968.

Kirk-Othmer, **9,** 138–148.

W. Kling: „Waschmittel 1928, 1968 und 2000," *Chem. Ind.* (*Düsseldorf*) **20** (1968) 394.

K. Durham: *Surface Activity and Detergency,* 1st ed., Macmillan, New York–London 1961.

H. Stüpel: *Synthetische Wasch- und Reinigungsmittel,* 2nd ed., Konradin-Verlag R. Kohlhammer, Stuttgart 1957.

R. Puchta, W. Grünewälder: *Textilpflege, Waschen und Chemischreinigen,* Verlag Schiele und Schön, Berlin 1973.

G. Wildbrett: *Technologie der Reinigung im Haushalt,* Verlag E. Ulmer, Stuttgart 1981.

Henkel & Cie, GmbH: *Waschmittelchemie,* Hüthig Verlag, Heidelberg 1976.

S. M. Mohnot: "A review of synthetic detergents," *CEW Chem. Eng. World* **10** (1975) 29.

H. Hauptmann: *Grundlagen der Wäschereichemie,* Spohr-Verlag, Frankfurt 1977, p. 451.

Surfactant Science Series, Marcel Dekker, New York. Vol. 1: M. Schick: *Nonionic Surfactants,* 1967.
Vol. 2: K. Shinoda: *Solvent Properties of Surfactant Solutions,* 1970.
Vol. 3: R. D. Swisher: *Surfactant Biodegradation,* 1970.

Vol. 4: E. Jungermann: *Cationic Surfactants,* 1970.
Vol. 5: W. G. Cutler, R. C. Davis: *Detergency: Theory and Test Methods,* 1972.
Vol. 6: K. J. Lissant: *Emulsions and Emulsion Technology,* 1974.
Vol. 7: W. M. Linfield: *Anionic Surfactants,* 1976.
Vol. 8: J. Cross: *Anionic Surfactants-Chemical Analysis,* 1977.
Vol. 9: T. Sato, R. Ruch: *Stabilization of Colloidal Dispersions by Polymer Adsorption,* 1979.
Vol. 10: Ch. Gloxhuber: *Anionic Surfactants, Biochemistry, Toxicology, Dermatology,* 1981.

M. Schick: *Nonionic Surfactants,* Marcel Dekker, New York 1967.

W. G. Cutler, R. C. Davis: *Detergency,* Part I–III, Marcel Dekker, New York 1973–1981.

„Chemie der Waschmittel und der Chemisch-Reinigung,“ *Chem. Ztg.* **99** (1975) 161–211.

K. Henning: „Aufbau, Funktion und Technologie moderner Waschmittel,“ *Chem. Labor Betr.* **27** (1976) 46, 81.

P. Berth: „Chemie und Technologie moderner Waschmittel,“ *Chem. Ztg.* **94** (1970) 974.

"Proceedings-World Conference on Soaps and Detergents," *J. Am. Oil Chem. Soc.* **55** (1978) 1–223.

McCutcheons Detergents & Emulsifiers, MC Publishing Co., Glen Rock, N.J. (lists producers and trade names, published yearly).

Arbeitsgemeinschaft der Verbraucherverbände (AGV), Institut für angewandte Verbraucherforschung (IfaV), *Info System Wäschepflege,* 1976, 1980.

H. Strone, W. Löbrich, R. Senf, K. D. Wetzler: *Bildungsfibel,* Part 1, Deutscher Textilreinigungs-Verband, Bonn 1980.

Henkel KGaA: *Industrielles Waschen,* Rheinisch-Bergische Druckerei GmbH, Düsseldorf 1978.

Grundlagen der Textilreinigung, VFB Fachbuchverlag, Leipzig 1974.

A. Davidsohn, B. M. Milwidsky: *Synthetic Detergents,* 6th ed., G. Godwin Ltd., London, J. Wiley & Sons, New York 1978.

A. Davidsohn: "Spray Drying and Dry Neutralization of Powdered Detergents," *J. Am. Oil Chem. Soc.* **55** (1978) 134.

J. Kretschmann: „Die großtechnische Herstellung konfektionierter Waschmittel — Mischen, Trocknen, Aufbereiten und Verpacken,“ *Hauswirtsch. Wissensch.* **16** (1968) 201.

G. Sorbe: „Verfahrenstechnik zur Herstellung phosphathaltiger Waschpulver,“ *Seifen Öle Fette Wachse* **105** (1979) 251.

M. J. Rosen, H. A. Goldsmith: *Systematic Analysis of Surface Active Agents,* 2nd ed., J. Wiley & Sons, London 1972.

D. Hummel: *Analyse der Tenside,* Hanser Verlag, München 1962.

R. Wickbold: *Die Analytik der Tenside,* Chem. Werke Hüls AG, Marl 1976.

B. M. Milwidsky: *Practical Detergent Analyses,* Mac Nair-Dorland Comp., New York 1970.

H. König: *Neuere Methoden zur Analyse von Tensiden,* Springer Verlag, Berlin 1971.

G. F. Longman: *The Analysis of Detergents and Detergent Products,* J. Wiley & Sons, London 1975.

H. Rath: *Lehrbuch der Textilchemie,* 3rd ed., Springer Verlag, Berlin 1972.

A. Chwala, V. Anger: *Handbuch der Textilhilfsmittel,* Verlag Chemie, Weinheim–New York 1977.

H. Reumuth: *Der Schmutz in seiner ganzen Vielfalt.* 100. Mitt. aus dem Institut für angewandte Mikroskopie, Photographie und Kinematographie-Karlsruhe der Fraunhofer-Gesellschaft e.V., 1965.

Specific References

[1] H. Bertsch, *Tenside* **5** (1968) 185–188.
[2] W. Kling, *Z. Gesamte Textilind.* **69** (1967) 86–93.
[3] Böhme Fettchemie, DE 640 997, 1928; DE 659 277, 1928.
[4] Henkel KGaA: *Fettalkohole, Rohstoffe, Verfahren und Verwendung,* Düsseldorf 1981.

[5] *Gesetz über Detergentien in Wasch- und Reinigungsmitteln,* 5.9.1961, BGBl. I, p. 1653.
[6] *Verordnung über die Abbaubarkeit von Detergentien in Wasch- und Reinigungsmitteln,* 1.12.1962, BGBl. I, pp. 698–706.
[7] G. Jakobi, M. J. Schwuger, *Chem. Ztg.* **99** (1975) 182–193.
[8] ISO 2174–1979 (E): *Surface active agents — Preparation of water with known calcium hardness.*
[9] B. Werdelmann, *Soap, Cosmet. Chem. Spec.* **50** (1974) 36.
[10] M. J. Schwuger in E. H. Lucassen-Reynders (ed.): *Anionic Surfactants,* Marcel Dekker, New York 1981.
[11] G. G. Jayson, *J. Appl. Chem.* **9** (1959) 422, 429.
[12] F. Jost, unpublished measurements.
[13] J. A. Finch, G. W. Smith in E. H. Lucassen-Reynders (ed.): *Anionic Surfactants,* Marcel Dekker, New York 1981.
[14] T. Young, *Philos. Trans. R. Soc. London* **95** (1805) 65, 82
[15] M. W. Fox, W. A. Zisman, *J. Colloid Sci.* **5** (1950) 514.
[16] E. G. Shafrin, W. A. Zisman, *J. Phys. Chem.* **64** (1960) 519.
[17] W. Kling, *Kolloid Z.* **115** (1949) 37.
[18] W. Kling, H. Lange, *Kolloid Z.* **142** (1955) 1.
[19] C. P. Kurzendörfer, H. Lange, M. J. Schwuger, *Ber. Bunsenges. Phys. Chem.* **82** (1978) 962.
[20] M. J. Schwuger, *J. Am. Oil Chem. Soc.* **59** (1982) 258, 265.
[21] M. J. Schwuger, H. G. Smolka, *Colloid Polym. Sci.* **255** (1977) 589.
[22] M. Saito, K. Shinoda, *J. Colloid Interface Sci.* **32** (1970) 647.
[23] C. P. Kurzendörfer, H. Lange, *Fette Seifen Anstrichm.* **71** (1969) 561.
[24] H. S. Kielman, P. J. F. van Steen in: *Surface Active Agents,* Society of Chemical Industry, London 1979.
[25] F. Jost, unpublished measurements.
[26] E. J. W. Verwey, J. T. G. Overbeek: *Theory of the Stability of Hydrophobic Colloids,* Elsevier, Amsterdam 1948.
[27] H. Sonntag, K. Strenge: *Koagulation und Stabilität disperser Systeme,* VEB Deutscher Verlag der Wissenschaften, Berlin (GDR) 1970.
[28] H. Lange in K. H. Mittal (ed.): *Adsorption at Interfaces,* ACS Symp. Ser. no. 8 (1975) 270.
[29] E. Hageböcke, *Dissertation,* Bonn 1956.
[30] M. J. Schwuger, *Ber. Bunsenges. Phys. Chem.* **83** (1979) 1193.
[31] D. Balzer, H. Lange, *Colloid Polym. Sci.* **253** (1975) 643; **255** (1977) 140.
[32] H. Lange, *Kolloid Z.* **169** (1960) 124; *J. Phys. Chem.* **64** (1960) 538.
[33] H. Lange in K. Shinoda (ed.): *Solvent Properties of Surfactant Solutions,* Marcel Dekker, New York 1967.
[34] P. Krings, M. J. Schwuger, C. H. Krauch, *Naturwissenschaften* **61** (1974) 75.
[35] M. J. Schwuger, H. G. Smolka, *Colloid Polym. Sci.* **254** (1976) 1062.
[36] H. G. Smolka, M. J. Schwuger, *Tenside Deterg.* **14** (1977) 222.
[37] H. G. Smolka, M. J. Schwuger, *Colloid Polym Sci.* **256** (1978) 270.
[38] M. J. Schwuger, H. G. Smolka, *Tenside Deterg.* **16** (1979) 233.
[39] R. Puchta, W. Grünewälder: *Textilpflege, Waschen und Chemischreinigen,* Verlag Schiele und Schön, Berlin 1973, pp. 47–100.
[40] M. J. Schwuger, *Chem. Ing. Tech.* **42** (1970) 433–438.
[41] H. Lange in K. Lindner (ed.): *Tenside-Textilhilfsmittel-Waschrohstoffe,* vol. 3, Wissenschaftl. Verlagsges., Stuttgart 1971, p. 2269.
[42] D. E. Haupt, *Soap, Cosmet. Chem. Spec.* **1984,** Sept., 42.
[43] G. Jakobi in H. Stache (ed.): *Tensid-Taschenbuch,* 2nd ed., Hanser Verlag, München-Wien 1981, pp. 253–337.

[44] M. F. Cox, T. P. Matson, J. L. Berna, A. Moreno, S. Kawakami, M. Suzuki, *J. Am. Oil Chem. Soc.* **61** (1984) no. 2, 330.

[45] H. Andree, P. Krings, *Chem. Ztg.* **99** (1975) 168–174.

[46] H. Andree, P. Krings in Henkel & Cie GmbH (ed.): *Waschmittelchemie,* Hüthig Verlag, Heidelberg 1976, pp. 73–90.

[47] M. J. Schwuger, *Chem. Ing. Tech.* **43** (1971) 705–710.

[48] M. J. Schwuger, *Fette Seifen Anstrichm.* **72** (1970) 565.

[49] IG-Farbenfabriken, DE 664 425, 1938.

[50] K. Bräuer, H. Fehr, R. Puchta, *Tenside Deterg.* **17** (1980) 281–287.

[51] W. P. Evans, *Chem. Ind.* (*London*) **1969,** 893.

[52] W. M. Linfield in E. Jungermann (ed.): *Surfactant Science Series, Cationic Surfactants,* vol. 4, Marcel Dekker, New York 1970, p. 49.

[53] H. W. Bücking, K. Lötzsch, G. Täuber, *Tenside Deterg.* **16** (1979) 2.

[54] W. Kling, *Münch. Beitr. Abwasser Fisch. Flußbiol.* **12** (1965) 38.

[55] G. C. Schweiker, *J. Am. Oil Chem. Soc.* **58** (1981) 58–61.

[56] G. Jakobi, M. J. Schwuger in Henkel & Cie GmbH (ed.): *Waschmittelchemie,* Hüthig Verlag Heidelberg 1976, pp. 91–120.

[57] A. E. Martell, G. Schwarzenbach, *Helv. Chim. Acta* **39** (1956) 653.

[58] B. Topley, *Q. Rev. Chem. Soc.* **3** (1949) 345.

[59] M. Becke-Goehring, H. Hoffmann: *Komplexchemie,* Springer Verlag, Heidelberg 1970, p. 50.

[60] W. Wirth, *Seifen Öle Fette Wachse* **96** (1970) 638.

[61] F. P. Dwyer, D. P. Mellor: *Chelating Agents and Metal Chelates.* Academic Press, New York 1964.

[62] F. J. Springer, *Vom Wasser* **34** (1967) 146; **35** (1968) 137.

[63] H. Bernhardt, W. Such, A. Wilhelms, *Münch. Beitr. Abwasser Fisch. Flußbiol,* **16** (1969) 60.

[64] P. Berth, G. Jakobi, E. Schmadel, *Chem. Ztg.* **95** (1971) 548–553.

[65] G. Zeit, *Chem. Ztg.* **96** (1972) 685–691.

[66] G. P. Lauhus, *Seifen Öle Fette Wachse* **98** (1972) 869–875.

[67] K. Enderer, K. Kosswig, *VI. Intern. Kongreß f. grenzflächenaktive Stoffe,* Zürich 1972.

[68] H. Haschke, G. Morlock, P. Kuzel, *Chem. Ztg.* **96** (1972) 199–207.

[69] K. Merkenich, K. Henning, H. Gudernatsch, *Seifen Öle Fette Wachse* **100** (1974) 433–439, 458–460.

[70] E. A. Matzner, M. M. Crutchfield, R. P. Langguth, R. D. Swisher, *Tenside Deterg.* **10** (1973) 119–125.

[71] E. A. Matzner, M. M. Crutchfield, R. P. Langguth, R. D. Swisher, *Tenside Deterg.* **10** (1973) 239–245.

[72] P. Berth, G. Jakobi, E. Schmadel, M. J. Schwuger, C. H. Krauch, *Angew. Chem.* **87** (1975) 115–123; *Angew. Chem. Int. Ed. Engl.* **14** (1975) 94.

[73] F. Smeets, R. van Oppen, E. Froyen, *Tenside Deterg.* **13** (1976) 83–89.

[74] V. Lamberti, *Meeting of the Chemical Specialties Manufactoring Association,* Chicago 1977.

[75] J. F. Schaffer, R. T. Woodhams, *Ind. Eng. Chem. Prod. Res. Dev.* **16** (1977) 3–11.

[76] Th. A. Downey, *Soap Chem. Spec.* **42** (1966) no. 2, 52–106.

[77] R. W. Cummins, *Deterg. Age* **5** (1968) no. 3, 22–27.

[78] Chem. Fabrik Budenheim, DE-OS 2 055 423, 1972.

[79] Colgate Palmolive, DE-OS 1 617 058, 1971.

[80] BASF, DE-OS 2 307 923, 1973.

[81] Henkel, DE-OS 2 412 837, 1974.

[82] Henkel, DE-OS 2 412 838, 1975.

[83] H. Lange, *Kolloid Z. Z. Polym.* **211** (1966) 106.

[84] G. Jakobi, *Angew. Makromol. Chem.* **123/124** (1984) 119.

[85] M. J. Schwuger, H. G. Smolka, *Colloid Polym. Sci.* **256** (1978) 1014–1020.
[86] H. Krüssmann, P. Vogel, H. Hloch, H. Carlhoff, *Seifen Öle Fette Wachse* **105** (1979) 3–6.
[87] M. Ettlinger, H. Ferch, *Seifen Öle Fette Wachse* **105** (1979) 131–135, 160.
[88] W. E. Adam, K. Neumann, J. P. Ploumen, *Fette Seifen Anstrichm.* **81** (1979) 445–449.
[89] M. J. Schwuger, H. G. Smolka, C. P. Kurzendörfer, *Tenside Deterg.* **13** (1976) 305–312.
[90] H. Nüsslein, K. Schumann, M. J. Schwuger, *Ber. Bunsenges. Phys. Chem.* **83** (1979) 1229–1238.
[91] C. P. Kurzendörfer, M. J. Schwuger, H. G. Smolka, *Tenside Deterg.* **16** (1979) 123–129.
[92] P. Berth, *Tenside Deterg.* **14** (1978) 176–180.
[93] H. Bloching, *Chem. Ztg.* **99** (1975) 194–201.
[94] D. Neher, H. Kniese, *Fette Seifen Anstrichm.* **72** (1970) 192–199.
[95] P. Kuzel, *Seifen Öle Fette Wachse* **105** (1979) 423–424.
[96] E. Rommel, *Seifen Öle Fette Wachse* **103** (1977) 411–416.
[97] P. Kuzel in J. Falbe (ed.): *Surfactants in Consumer Products,* Springer Verlag, Heidelberg 1987, p. 267.
[98] H. Kauffmann, *Angew. Chem.* **37** (1924) 364; **43** (1930) 840.
[99] A. Agster, *Textilveredlung* **1** (1966) 276.
[100] I. E. Flis, *Tr. Leningr. Tekhnol. Inst. Tsellyulozno-Bumazhn. Prom.* **1964,** no. 12, 37–49.
[101] H. Bloching, *Sonderheft „Haushaltswaschmittel"*, Ciba-Geigy, 1975, p. 17.
[102] D. Coons in J. Falbe (ed.): *Surfactants in Consumer Products,* Springer Verlag, Heidelberg 1987, p. 274.
[103] H. Milster, *Seifen Öle Fette Wachse* **97** (1971) 549–550.
[104] G. Jakobi in J. Falbe (ed.): *Surfactants in Consumer Products,* Springer Verlag, Heidelberg 1987, pp. 214, 286.
[105] Unilever, DE 1 291 317, 1969.
[106] Henkel, DE 1 594 865, 1969; DE 1 695 219, 1972.
[107] H. Goy, *Seifen Öle Fette Wachse* **100** (1974) 401–404, 451–452.
[108] Noury van der Lande, DE 1 162 967, 1960.
[109] J. Mecheels, *Seifen Öle Fette Wachse* **108** (1982) 31–34.
[110] Procter & Gamble, US 4 412 934, 1983.
[111] A. Gilbert, *Deterg. Age* **4** (1967) July, 30–33.
[112] A. Gilbert, *Deterg. Age* **4** (1967) June, 18–20.
[113] W. Trieselt, *Melliand Textilber. Int.* **51** (1970) no. 9, 1094–1097.
[114] Henkel, DE 703 604, 1933.
[115] O. Röhm, DE 283 923, 1915.
[116] O. Koch, *Seifen Öle Fette Wachse* **95** (1969) 663–666.
[117] M. Berg, A. Boeck in Henkel & Cie GmbH (ed.): *Waschmittelchemie,* Hüthig Verlag, Heidelberg 1976, pp. 155–178.
[118] H. Andree, W. R. Müller, R. D Schmid, *J. Appl. Biochem.* **2** (1980) 218.
[119] G. Jakobi, *Tenside* **6** (1969) 307–311.
[120] J. W. Hensley, *J. Am. Oil Chem. Soc.* **42** (1965) 993–997.
[121] J. Stawitz, P. Höpfner, *Seifen Öle Fette Wachse* **86** (1960) 51–52.
[122] P. G. Evans, W. P. Evans, *J. Appl. Chem.* **17** (1967) 276–282.
[123] G. K. Greminger Jr., *Soap Cosmet. Chem. Spec.* **2** (1978) Nov., 28–31, 38.
[124] G. K. Greminger, *J. Am. Oil Chem. Soc.* **55** (1978) 122–126.
[125] BASF, DE 806 366, 1948.
[126] Colgate Palmolive, CH 300 865, 1950.
[127] Procter & Gamble, CH 314 599, 1951.
[128] Henkel, DE-AS 1 135 606, 1961.
[129] H. Y. Lew, *J. Am. Oil Chem. Soc.* **41** (1964) 297–300.

[130] E. Schmadel, *Fette Seifen Anstrichm.* **70** (1968) 491.
[131] Henkel, DE 1 257 338, 1968.
[132] J. Perner, G. Frey, K. Stork, *Tenside Deterg.* **14** (1977) 180–185.
[133] H. Distler, D. Stoeckigt, *Tenside Deterg.* **12** (1975) 263–265.
[134] Procter & Gamble, DE 1 056 316, 1966.
[135] Procter & Gamble, DE-AS 1 080 250, 1960.
[136] Procter & Gamble, DE-OS 2 338 464, 1974.
[137] Dow Corning, DE-OS 2 402 955, 1974.
[138] G. Rossmy, *Fette Seifen Anstrichm.* **71** (1969) 56.
[139] H. Gold, *MVC-Report 2. Fluorescent Whitening Agents,* Stockholm 1973.
[140] V. DiGiacomo, *Soap Chem. Spec.* **1** (1967) Jan., 79–82, 119.
[141] E. Obrocki, *Fette Seifen Anstrichm.* **73** (1971) 327–329.
[142] W. Sturm, G. Mansfeld, *Tenside Deterg,* **15** (1978) 181–186.
[143] B. Streschnak, *Parfüm. Kosm.* **61** (1980) 285–289.
[144] P. L. Layman, *Chem. Eng. News* **1984,** Jan., 23, 31.
[145] D. H. Scharer, H. Stupel, J. G. Moffet Jr., *Soap Cosmet. Chem. Spec.* **1** (1974) no. 4, 33–36, 53.
[146] D. E. Haupt, *Soap Cosmet. Chem. Spec.,* **1984,** April, 31.
[147] H. Andree, P. Krings, H. Verbeek, *Seifen Öle Fette Wachse* **107** (1981) 115–119.
[148] E. Sung in J. Falbe (ed.): *Surfactants in Consumer Products,* Springer Verlag, Heidelberg 1987, pp. 292–295.
[149] D. M. Coons, *J. Am. Oil Chem. Soc.* **55** (1978) Jan., 104.
[150] R. Puchta, *Seifen Öle Fette Wachse* **104** (1978) 177–182.
[151] R. Puchta, *J. Am. Oil Chem. Soc.* **61** (1984) Feb., 367–376.
[152] R. R. Egan, *J. Am. Oil Chem. Soc.* **55** (1978) 118–121.
[153] W. Mooney, *Text. Mon.* **1980,** Oct., 32–34, 68–71.
[154] R. B. McConnell, *Am. Dyest. Rep.* **67** (1978) July, 31–34.
[155] H. W. Bücking, K. Lötzsch, G. Täuber, *Tenside Deterg.* **16** (1979) 1–10.
[156] L. Hughes, J. M. Leiby, M. L. Deviney, *Soap Cosmet. Chem. Spec.* **2** (1975) Oct., 56–62.
[157] J. H. Barrett, *Household Pers. Prod. Ind.* **2** (1980) Sept., 58–65; **2** (1980) Oct., 42–46.
[158] *Consum. Rep.* **1979,** Jan., 48–49.
[159] R. J. De Vries, *SPC Soap Perfum. Cosmet.* **50** (1977) June, 223–224, 227.
[160] H. Grund, *Wäscherei Tech. Chem.* **22** (1969) 378.
[161] W. Bechstedt, W. A. Roland, *Reiniger Wäscher* **33** (1980) no. 6, 21–29, no. 7, 18–24.
[162] B. Ziolkowsky et al., *Seifen Fette Öle Wachse* **107** (1981) 333.
[163] K. Henning, *Seifen Fette Öle Wachse* **104** (1978) 407.
[164] G. A. Krause, DE 297 388, 1912.
[165] A. P. Lamont, US 51 263, 1865.
[166] R. L. Holliday, US 1 621 506, 1927.
[167] A. P. Lamont, US 1 652 900, 1927.
[168] W. R. Marshall Jr., "Atomization and Spray Drying," *Chem. Eng. Progress Monograph* **50** (1954) no. 2.
[169] D. W. Belcher, D A. Smith, E. M. Cook, *Chem. Eng.* (*London*) **70** (1963) Oct. 14, 201–208.
[170] H. G. Kessler, *Chem. Ing. Tech.* **39** (1967) 601–606.
[171] W. Armbruster, *Chem. Ztg.* **93** (1969) 493–504.
[172] C. W. Lyne, *Br. Chem. Eng.* **16** (1971) 370–371.
[173] O. T. Kragh, A. Kraglund, *Chem. Eng.* (*London*) **367** (1981) April, 149–153.
[174] J. H. Chaloud, J. B. Martin, J. S. Baker, *Chem. Eng. Prog.* **53** (1957) 593.
[175] VDI-Bildungswerk, Düsseldorf: *Prozeßrechner in der Verfahrenstechnik.*
[176] G. Pajek, K. Kurth: *Stetigförderer,* VEB, Verlag Technik, Berlin (GDR) 1967.

[177] H. J. Taubmann, F. Bauer, *Aufbereit. Tech.* **12,** (1971) 466.
[178] J. Schwedes: *Fließverhalten von Schüttgütern in Bunkern*, Verlag Chemie, Weinheim 1968.
[179] J. Schwedes, *Aufbereit. Tech.* **10** (1969) 533.
[180] F. W. Plank, *Zem. Kalk Gips* **27** (1974) 271.
[181] W. Müller, *Chem. Ing. Tech.* **53** (1981) 831.
[182] H. E. Tschakert, *Seifen Öle Fette Wachse* **98** (1972) 607, 661.
[183] K. Merkenich, K. Henning, *Seifen Öle Fette Wachse* **99** (1973) 351.
[184] O. Pfrengle, M. Beckmann, *Seifen Öle Fette Wachse* **99** (1973) 358.
[185] H. G. Smolka: *Betrachtungen der Energieumsätze auf dem Wasch-, Spül- und Reinigungsmittelsektor mit dem Ziel der Energieeinsparung*, AIS-Meeting, Reykjavik/Island, Sept. 1981.
[186] J. Kretschmann, H. Latka, H. Reuter, in Henkel & Cie GmbH (ed.): *Waschmittelchemie*, Hüthig Verlag, Heidelberg 1976, pp. 232–252.
[187] H. Hoffmann, G. Hohlfeld, J. M. Quack, *Seifen Öle Fette Wachse* **104** (1978) 209.
[188] W. E. Adam, K. Neumann, *Fette Seifen Anstrichm.* **80** (1978) 392.
[189] H. D. Nielen, H. Landgräber, *Tenside Deterg.* **14** (1977) 205.
[190] G. Sorbe, *Seifen Öle Fette Wachse* **105** (1979) 251.
[191] Ullmann 4th ed., **13,** 292–294; **17,** 9–18.
[192] M. Ettlinger, H. Ferch, *Seifen Öle Fette Wachse* **105** (1979) 131.
[193] H. Strack, *Seifen Öle Fette Wachse* **105** (1979) 457.
[194] W. E. Adam, K. Neumann, J. P. Ploumen, *Fette Seifen Anstrichm.* **81** (1979) 445.
[195] Union Carbide Corp., DE 1 038 017, 1954.
Martinswerk, DE 1 667 620, 1967.
Degussa, DE 2 333 068, 1973.
Degussa, Henkel, DE 2 447 021, 1974; DE 2 514 399, 1975; DE 2 517 218, 1975; DE 2 651 485, 1976; DE 2 651 445, 1976; DE 2 651 436, 1976
DE 2 651 419, 1976; DE 2 651 420, 1976; DE 2 651 437, 1976; DE 2 704 310, 1977;
DE 2 734 296, 1977; DE 2 856 278, 1978; DE 2 910 147, 1979; DE 2 910 152, 1979;
DE 2 941 636, 1979; DE 3 007 044, 1980; DE 3 007 080, 1980; DE 3 007 087, 1980;
DE 3 007 123, 1980; DE 3 011 834, 1980; DE 3 021 370, 1980.
J. M. Huber Corp., DE 2 633 304, 1976, DE 2 951 192, 1979.
Montedison, DE 3 002 278, 1980.
[196] Henkel, DE 2 412 838, 1974; DE 2 527 388, 1975; DE 2 615 698, 1976; DE 2 702 979, 1977.
[197] O. Koch, *Seifen Öle Fette Wachse* **106** (1980) 321.
[198] O. Koch, *Fette Seifen Anstrichm.* **75** (1973) 331.
[199] H. J. Lehmann, *Chem. Unserer Zeit* **7** (1973) 82–89.
[200] E. P. Frieser, *Österr. Chem. Z.* **76** (1975) no. 9, 4–8; **76** (1975) no. 10, 1–6; **76** (1975) no. 11, 5–7; **76** (1975) no. 12, 5–9.
[201] K. Henning, *Chem. Lab. Betr.* **27** (1976) 46–50, 81–86.
[202] G. Gawalek: *Tenside*, Akademie Verlag, Berlin 1975.
[203] Henkel: *Waschmittelchemie*, Hüthig Verlag, Heidelberg 1976.
[204] P. Berth, G. Jakobi, E. Schmadel, J. Schwuger, C. Krauch, *Angew. Chem.* **87** (1975) 115–123; *Angew. Chem. Int. Ed. Engl.* **14** (1975) 94.
[205] G. Sorbe, *Seifen Öle Fette Wachse* **105** (1979) 251–253.
[206] W. M. Linfield: *Anionic Surfactants*, vol. 7 (Parts I and II), Marcel Dekker, New York-Basel 1976.
[207] P. Berth, M. J. Schwuger, *Tenside Deterg.* **16** (1979) 175–184.
[208] P. Berth, J. Heidrich, G. Jakobi, *Tenside Deterg.* **17** (1980) 228–235.
[209] B. Ziolkowsky, *Seifen Öle Fette Wachse* **12** (1981) 333–336, 339–343.
[210] Th. Kunzmann, *Seifen Öle Fette Wachse* **98** (1972) 179–181.
[211] M. J. Schwuger, H. G. Smolka, *Colloid Polym. Sci.* **254** (1976) 1062.

[212] M. J. Schwuger, H. G. Smolka, C. P. Kurzendörfer, *Tenside Deterg.* **13** (1976) 305.
[213] C. P Kurzendörfer, M. J. Schwuger, H. G. Smolka, *Tenside Deterg.* **16** (1979) 123–129.
[214] H. G. Smolka, M. J. Schwuger, *Tenside Deterg.* **14** (1977) 222.
[215] H. G. Smolka, M. J. Schwuger, *Colloid Polymer. Sci.* **256** (1978) 270–277.
[216] M. J Schwuger, H. G. Smolka, *ACS Symp. Ser.* **40** (1971) 696–707.
[217] M. J. Schwuger, H. G. Smolka, *Tenside Deterg.* **16** (1979) 233–239.
[218] W. Kling, *Angew. Chem.* **65** (1953) 201–212.
[219] W. Kling, *Parfüm. Kosmet.* **45** (1964) 1–5, 29–31.
[220] W. Hagge, *CID-Kongreß 1968,* vol. I, pp. 1–19.
[221] G. F. Longman: *Analysis of Detergents and Detergent Products,* J. Wiley & Sons, London 1975.
[222] N. Schönfeld: *Grenzflächenaktive Ethylenoxid-Addukte,* Wissenschaftl. Verlagsges., Stuttgart 1976, pp. 991–1045.
[223] J. Wiesner, L. Wiesnerova, *J. Chromatogr.* **114** (1975) 411–412.
[224] E. Kunker, *Tenside Deterg.* **18** (1981) 301–305.
[225] E. Heinerth, *Seifen Öle Fette Wachse* **5** (1964) 105–110.
[226] R. Wickbold: *Die Analytik der Tenside,* Chemische Werke Hüls AG, Marl 1976.
[227] G. Graffmann, W. Hörig, P. Sladek, *Tenside Deterg.* **14** (1977) 194–197.
[228] M. Teupel, *CID-Kongreß 1960,* vol. C/III, pp. 177–183.
[229] G. Schwarz, *Seifen Öle Fette Wachse* **22** (1961) 715–717.
[230] H. König, *Fresenius' Z. Anal. Chem.* **293** (1978) 295–300.
[231] M. J. Rosen, H. A. Goldsmith: *Sytematic Analysis of Surface-Active Agents,* 2nd ed., J. Wiley & Sons, London 1972.
[232] Schweizerische Gesellschaft f. Analytische und Angewandte Chemie: *Seifen und Waschmittel,* Verlag H. Huber, Bonn-Stuttgart 1955, 1963.
[233] D. Hummel: *Analyse der Tenside,* Hanser Verlag, München 1962.
[234] B. M. Milwidsky, *Practical Detergent Analyses,* Mac Nair-Dorland Comp., New York 1970.
[235] British Standard 3762:1964, "Methods of Sampling and Testing Detergents."
[236] G. F. Longman, *Talanta* **22** (1975) 621–636.
[237] *Jahrbuch für den Praktiker aus der Öl-, Fett-, Seifen- und Waschmittel-, Wachs- und sonstigen chemischtechnischen Industrie 1981,* Verlag für chemische Industrie H. Ziolkowsky KG, Augsburg.
[238] J. Puschmann, *Angew. Makromol. Chem.* **47** (1975) 29–41.
[239] H. Specker, *Angew. Chem.* **80** (1968) 297–304; *Angew. Chem. Int. Ed. Engl.* **7** (1968) 252.
[240] G. Tölg, *Naturwissenschaften* **63** (1976) 99–110.
[241] DGF-Einheitsmethoden, Wissenschaftl. Verlagsges. Stuttgart 1950–1984.
[242] Official and Tentative Methods of The American Oil Chemists Society, 508 South Sixth Street, Champaign, Illinois 61 820.
[243] ISO-TC 91.
[244] G. Graffmann, H. Domels, W. Hörig, *Fette Seifen Anstrichm.* **77** (1975) 364–365.
[245] DIN Deutsches Institut für Normung e.V., Beuth-Vertrieb GmbH, Berlin-Köln.
[246] Jander-Blasius: *Lehrbuch der analytischen und präparativen anorganischen Chemie,* Hirzel Verlag, Stuttgart 1973.
[247] Anorganikum, VEB Deutscher Verlag der Wissenschaften, Berlin (GDR) 1967.
[248] Fresenius-Jander: *Handbuch der Analytischen Chemie,* Springer Verlag, Berlin 1953.
[249] Deutsche Einheitsverfahren zur Wasseruntersuchung, Verlag Chemie, Weinheim 1960 ff.
[250] H. Puderbach, *J. Am. Oil Chem. Soc.* **55** (1978) 156–162.
[251] I. Kawase, A. Nakae, K. Tsuji, *Anal. Chim. Acta* **131** (1981) 213–222.
[252] F. Feigl: *Spot Tests in Organic Analysis,* 7th ed., Elsevier, Amsterdam 1966.
[253] F. Cramer: *Papierchromatographie.* 5th ed., Verlag Chemie, Weinheim 1962.

[254] E. Stahl: *Dünnschichtchromatographie*. 2nd ed., Springer Verlag, Berlin 1967.
[255] E. Heinerth, J. Pollerberg, *Fette Seifen Anstrichm.* **61** (1959) 376–377.
[256] R. Pribil, V. Vesely, *Chem. Analyst* **56** (1967) 51–53.
[257] E. Heinerth, *Fette Seifen Anstrichm.* **70** (1968) 495–498.
[258] J. Drewry, *Analyst (London)* **88** (1963) 225–231.
[259] D. Hummel: *Kunststoff-, Lack- und Gummianalyse,* vol. I, Hanser Verlag, München 1958.
[260] E. Knappe, I. Rhodewald, *Fresenius' Z. Anal. Chem.* **223** (1966) 174–181.
[261] K. Figge, *Fette Seifen Anstrichm.* **70** (1968) 680–687.
[262] G. Lehmann, M. Becker-Klose, *Tenside Deterg.* **13** (1976) 7–8.
[263] H. Bloching, W. Holtmann, M. Otten, *Seifen Öle Fette Wachse* **105** (1979) 33–38, 82–83.
[264] S. R. Epton, *Nature (London)* **160** (1947) 795–796.
[265] R. Matissek, E. Hieke, W. Baltes, *Fresenius' Z. Anal. Chem.* **300** (1980) 403–406.
[266] H. Brüschweiler, V. Sieber, H. Weishaupt, *Tenside Deterg.* **17** (1980) 126–129.
[267] M. Däuble, *Tenside Deterg.* **18** (1981) 7–12.
[268] J. M. Rosen, *Anal. Chem.* **27** (1955) 787–790.
[269] K. Bürger, *Fresenius' Z. Anal. Chem.* **196** (1963) 251–259.
[270] *Erfahrungsaustausch der Seifen-, Wasch- und Reinigungsmittelindustrie*, vol. 1, pp. 9–12, 1944; Vagda-Kalender, 5th ed., 1943, p. 87.
[271] A. Hintermaier, *Fette Seifen* **1944,** Sept. 1, 367–368.
[272] P. Friese, *Fresenius' Z. Anal. Chem.* **303** (1980) 279–288.
[273] H. P. Kaufmann: *Analyse der Fette und Fettprodukte,* Springer Verlag, Berlin 1958.
[274] R. Wickbold, *Tenside Deterg.* **9** (1972) 173–178.
[275] H. Hellmann, *Fresenius' Z. Anal. Chem.* **300** (1980) 44–47.
[276] B. Weibull: *3. Int. Kongreß für grenzflächenaktive Stoffe, Köln 1960,* vol. 3, Universitätsdruckerei, Mainz 1961, p. 121.
[277] W. Riemann, H. Walton: *Ion Exchange in Analytical Chemistry,* Pergamon Press, Oxford 1970.
[278] P. Voogt, *Recl. Trav. Chim. Pays-Bas* **77** (1958) 889–901.
[279] R. Wickbold, *Seifen Öle Fette Wachse* **86** (1960) 79–82.
[280] N. Blumer, *Schweiz. Arch.* **29** (1963) 171–180.
[281] M. E. Ginn, C. C. Church, *Anal. Chem.* **31** (1959) 551–555.
[282] K. Bey, *Fette Seifen Anstrichm.* **67** (1965) 25–30.
[283] P. Voogt in H. A. Boeckenoogen (ed.): *Analysis and Characterization of Oils, Fats, and Fat Products* Interscience Publ. Inc., London 1964.
[284] R. Wickbold, *Tenside Deterg.* **13** (1976) 177–180.
[285] R. Wickbold, *Tenside Deterg.* **13** (1976) 181–187.
[286] G. Schwarz, *Fette Seifen Anstrichm.* **71** (1969) 223–226.
[287] E. Heinerth, *Fette Seifen Anstrichm.* **63** (1961) 181–183.
[288] R. Bock: *Aufschlußmethoden der anorg. und organischen Chemie,* Verlag Chemie, Weinheim 1972.
[289] Verein Deutscher Eisenhüttenleute: *Handbuch für das Eisenhüttenlaboratorium,* vol. 1, Stahleisen-Verlag, Düsseldorf 1960.
[289 a] *Fette Seifen Anstrichm.* **79** (1977) 203.
[290] G. Staats, H. Brück, *Fresenius' Z. Anal. Chem.* **250** (1970) 289–294.
[291] E. Vaeth, E. Griessmayer, *Fresenius' Z. Anal. Chem.* **303** (1980) 268–271.
[292] O. G. Koch, A. Koch-Dedic: *Handbuch der Spurenanalyse,* Springer Verlag, Berlin 1974.
[293] J. Drozd, *J. Chromatogr.* **113** (1975) 303–356.
[293 a] *Fette Seifen Anstrichm.* **74** (1972) 31.
[294] A. Hintermaier, *Angew. Chem.* **60** (1948) 158–159.
[295] V. W. Reid, G. F. Longman, E. Heinerth, *Tenside* **4** (1967) 292–304.

[296] S. Lee, N. A. Puttnam, *J. Am. Oil Chem. Soc.* **43** (1966) 690.
[297] R. Denig, *Fette Seifen Anstrichm.* **76** (1974) 412–416.
[298] R. Denig, *Tenside Deterg.* **10** (1973) 59–63.
[299] W. Kupfer, K. Künzler, *Fresenius' Z. Anal. Chem.* **267** (1973) 166–169.
[300] T. H. Liddicoet, L. H. Smithson, *J. Am. Oil Chem. Soc.* **42** (1965) 1097–1102.
[301] ASTM Standards of Soaps and other Detergents, Part 30, D 501–67, 1974.
[302] A. J. Sheppard, J. L. Iversen, *J. Chromatogr. Sci.* **13** (1975) 448–452.
[303] H. Schlenk, J. L. Gellerman, *Anal. Chem.* **32** (1960) 1412–1414.
[304] L. D. Metcalf, A. A. Schmitz, *Anal. Chem.* **33** (1961) 363–364.
[305] A. Seher, H. Pardun, M. Arens, *Fette Seifen Anstrichm.* **80** (1978) 58–66.
[306] J. D. Knight, R. House, *J. Am. Oil Chem. Soc.* **36** (1959) 195–200.
[307] J. Pollerberg, *Fette Seifen Anstrichm.* **67** (1965) 927–929.
[308] M. J. Rosen, G. C. Goldfinger, *Anal. Chem.* **28** (1956) 1979–1981.
[309] J. Borecky, *Mikrochim. Acta* **1962,** 1137–1145.
[310] S. Nishi, *Bunseki Kagaku* **14** (1965) 917; *Chem. Abstr.* **64** (1966) 3858f.
[311] R. Wickbold, *Tenside Deterg.* **12** (1975) 25–27.
[312] *Houben-Weyl,* **8** „Sauerstoffverbindungen III."
[313] H. Biltz, W. Biltz: *Ausführung quantitativer Analysen.* Hirzel Verlag, Stuttgart 1965.
[314] British Standard 3984: 1966 "Sodium Silicates".
[315] G. Graffmann, W. Schneider, L. Dinkloh, *Fresenius' Z. Anal. Chem.* **301** (1980) 364–372.
[316] J. Longwell, W. D. Maniece, *Analyst* (*London*) **80** (1955) 167–171.
[317] M. Kolthoff, V. A. Stenger: *Volumetric Analysis,* 2nd ed., vols I and II, Interscience Publ., New York 1942, 1947.
[318] G. Jander, K. F. Jahr, H. Knoll: *Maßanalyse,* De Gruyter, Berlin 1966.
[319] E. Heinerth, *Tenside* **4** (1967) 45–47.
[320] S. Ebel, U. Parzefall: *Experimentelle Einführung in die Potentiometrie,* Verlag Chemie, Weinheim 1975.
[321] H. Malmerig, *GIT Fachz. Lab.* **19** (1975) 400–404.
[322] S. Ebel, A. Seuring, *Angew. Chem.* **89** (1977) 129–141; *Angew. Chem. Int. Ed. Engl.* **16** (1977) 157.
[323] A. Hofer, E. Brosche, R. Heidinger, *Fresenius' Z. Anal. Chem.* **253** (1971) 117–119.
[324] J. Koryta, *Anal. Chim. Acta* **61** (1972) 329–411.
[325] *Orion Ion- and Gas-sensitive Electrodes*, Orion Research Inc., Cambridge 1973.
[326] G. Schwarzenbach, H. Flaschka: *Die komplexometrische Titration,* Enke Verlag, Stuttgart 1965.
[327] R. Pribil: *Komplexometrie,* VEB Deutscher Verlag für Grundstoffind., Leipzig 1960.
[328] G. Graffmann, H. Domels, M. L. Sträter, *Fette Seifen Anstrichm.* **76** (1974) 218–220.
[329] Merck AG: *Komplexometrische Bestimmungen mit Titriplex,* 3rd ed., Darmstadt 1966.
[330] R. Wickbold: *VI. Int. Kongreß für grenzflächenaktive Stoffe, Zürich 1972*, vol. I, Section A, Hanser Verlag, München 1973, p. 373.
[331] F. Öhme: *Angewandte Konduktometrie,* Hüthig Verlag, Heidelberg 1962.
[332] K. Kiemstedt, W. Pfab, *Fresenius' Z. Anal. Chem.* **213** (1965) 100–107.
[333] J. Cross: *Anionic Surfactants — Chemical Analysis,* vol. 8, Marcel Dekker, New York-Basel 1977.
[333a] *Fette Seifen Anstrichm.* **73** (1971) 683.
[334] H. R. Hoffmann, W. Böer, G. W. G. Schwarz, *Fette Seifen Anstrichm.* **78** (1976) 367–368.
[335] *Pharmacopée Française* 9th ed., "Benzalkonium (Chlorure de)," Adapharus, Paris 1972.
[336] R. Wickbold, *Seifen Öle Fette Wachse* **85** (1959) 415–416.
[337] D. C. White, *Mikrochim. Acta* **1959,** 254–259.
[338] E. E. Archer, *Analyst* (*London*) **82** (1957) 208–209.

[339] AFNOR, Doc. Française A-124 (7.7.70)
[340] R. E. Kitson, M. G. Mellon, *Ind. Eng. Chem. Anal. Ed.* **16** (1944) 379–383.
[341] H. Bernhardt, (Ausschuß Phosphate und Wasser), *Zeitschr. f. Wasser Abwasser Forsch.* **7** (1974) 143–146.
[342] G. van Raay, *Tenside Deterg.* **7** (1970) 125–132.
[343] H. Jaag: *Über proteolytische Enzyme, deren Prüfmethoden und Einsatzmöglichkeiten in der Waschmittelindustrie,* Verlag H. Lang & Cie AG, Bern 1968.
[344] Boehringer, Mannheim: *Methoden der enzymatischen Lebensmittelanalytik,* 1980.
[345] L. K. Wang, *J. Am. Oil Chem. Soc.* **52** (1975) 339–344.
[346] K. Toel, H. Fujii, *Anal. Chim. Acta* **90** (1977) 319–322.
[347] St. Janeva, R. Borissova-Pangarova, *Talanta* **25** (1977) 279–282.
[348] W. Kupfer, *Tenside Deterg.* **12** (1975) 40.
[349] R. Hermann, C. Th. J. Alkemade: *Flammenphotometrie,* Springer Verlag, Berlin-Göttingen-Heidelberg 1960.
[350] B. Welz: *Atom-Absorptions-Spektroskopie,* Verlag Chemie, Weinheim 1972.
[351] H. Bernd, E. Jackwerth, *Spectrochim. Acta Part B* **30 B** (1975) 169–177.
[352] G. Volland, G. Kölblin, P. Tschöpel, G. Tölg, *Fresenius' Z. Anal. Chem.* **284** (1977) 1–12.
[353] E. Jackwerth, J. Lohmar, G. Wittler, *Fresenius' Z. Anal. Chem.* **266** (1973) 1–8.
[354] A. Dornemann, H. Kleist, *Fresenius' Z. Anal. Chem.* **291** (1978) 353–359.
[355] R. O. Müller: *Spektrochemische Analyse mit Röntgenfluoreszenz,* R. Oldenbourg Verlag, München 1967.
[356] E. Vaeth, E. Griessmayr, *Fresenius' Z. Anal. Chem.* **303** (1980) 268–271.
[357] E. Schenbeck, Ch. Jörrens, *Fresenius' Z. Anal. Chem.* **303** (1980) 257–264.
[358] H. Krischner: *Einführung in die Röntgenfeinstrukturanalyse,* Vieweg & Sohn, Braunschweig 1974.
[359] J. M. Mabis, O. T. Quimby, *Anal. Chem.* **25** (1953) 1814–1818.
[360] D. W. Breck: *Zeolite Molecular Sieves: Structure, Chemistry and Use,* J. Wiley & Sons, New York 1974, pp. 347–378.
[361] K. Lötzsch, *Seifen Öle Fette Wachse* **105** (1979) 261–267.
[362] G. Zweig, J. Sherma: *Handbook of Chromatography,* CRC-Press, Cleveland, Ohio, 1972.
[363] F. Korte: *Methodicum Chimicum,* vol. 1: *Analyticum,* Part 1, G. Thieme Verlag, Stuttgart 1973.
[364] J. M. Hais, K. Macek: *Handbuch der Papierchromatographie,* VEB Fischer Verlag, Jena 1963.
[365] H. Grunze, E. Thilo: *Die Papierchromatographie der kondensierten Phosphate,* Sitzungsberichte der Deutschen Akademie der Wissenschaften zu Berlin, Akademie Verlag, Berlin 1955.
[366] E. Heinerth, J. Pollerberg, *Fette Seifen Anstrichm.* **61** (1959) 376–377.
[367] R. E. Cline, R. M. Fink, *Anal. Chem.* **28** (1956) 47–52.
[368] J. C. Touchstone, M. F. Dobbins: *Practice of the Thin Layer Chromatography,* J. Wiley & Sons, New York 1978.
[369] F. Geiss: *Die Parameter der DC,* Vieweg und Sohn, Braunschweig 1972.
[370] E. Heinerth, *Fette Seifen Anstrichm.* **70** (1968) 495–498.
[371] A. Breyer, M. Fischl, E. Setzer, *J. Chromatogr.* **82** (1973) 37–52.
[372] K. Bey, *Fette Seifen Anstrichm.* **67** (1965) 217–221.
[373] U. Hezel, *Angew. Chem.* **85** (1973) 334–342; *Angew. Chem. Int. Ed. Engl.* **12** (1973) 298.
[374] J. Touchstone, T. Murawec, M. Kasparow, W. Wortmann, *J. Chromatogr. Sci.* **10** (1972) 490–493.
[375] D. H. Liem, *Cosmet. Toiletries* **92** (1977) 59–72.
[376] H. König, *Fresenius' Z. Anal. Chem.* **251** (1970) 359–368.
[377] R. Matissek, E. Hieke, U. Baltes, *Fresenius' Z. Anal. Chem.* **300** (1980) 403–406.
[378] M. Köhler, B. Chalupka, *Fette Seifen Anstrichm.* **84** (1982) 208–211.

[379] R. Matissek, *Tenside Deterg.* **19** (1982) 57–66.
[380] S. J. Patterson, E. C. Hunt, K. B. E. Tucker, *J. Proc. Inst. Sewage Purif.* **1966,** 190.
[381] G. Hesse: *Chromatographisches Praktikum,* Akademische Verlagsges., Frankfurt 1968.
[382] K. Ehlert, R. Engler, *GIT Fachz. Lab.* **23** (1979) 659–664.
[383] P. Quinlin, H. J. Weiser, *J. Am. Oil Chem. Soc.* **35** (1958) 325–327.
[384] R. Wickbold, *Fette Seifen Anstrichm.* **74** (1972) 578–579.
[385] W. Kupfer, K. Künzler, *Fette Seifen Anstrichm.* **74** (1972) 287–291.
[386] M. J. Rosen, *Anal. Chem.* **35** (1963) 2074–2077.
[387] R. E. Kaiser: *Chromatographie in der Gasphase,* Bibliographisches Institut, Mannheim 1973.
[388] G. Schomburg, R. Dielmann, H. Husmann, F. Weeke, *J. Chromatogr.* **122** (1976) 55–72.
[389] H. J. Vonk et al.: *Int. Surfactants Congress, Moscow 1976,* vol. 1, Section A, pp. 435–449.
[390] G. Schomburg: *Gaschromatographie,* Verlag Chemie, Weinheim 1977.
[391] A. Kuksis, *Fette Seifen Anstrichm.* **73** (1971) 130–138.
[392] H. Hadorn, K. Zürcher, *Dtsch. Lebensm. Rundsch.* **66** (1970) 77–87.
[393] A. F. Prevot, F. X. Mordret, *Rev. Fr. Corps Gras* **23** (1976) 409–423.
[394] W. J. Carnes, *Anal. Chem.* **36** (1964) 1197–1200.
[395] E. Link, H. M. Hickman, R. A. Morrissette, *J. Am. Oil Chem. Soc.* **36** (1959) 20–23, 300–303.
[396] D. G. Anderson, *J. Paint. Technol.* **40** (1968) 549–557.
[397] B. Bonney, N. F. Glennard, A. M. Humphrey, *Chem. Ind.* (*London*) **1973,** 749–751.
[398] R. Schicker, *Aerosol Rep.* **13** (1974) 149–163.
[399] R. Schubert, L. Ketel, *J. Soc. Cosmet. Chem.* **23** (1972) 115–124.
[400] Hoechst, DE-OS 2 161 702, 1971.
[401] T. J. Birkel, C. R. Warner, T. Fario, *J. Am. Oil Chem. Soc.* **62** (1979) 931–936.
[402] H. Engelhardt: *Hochdruck-Flüssigkeits-Chromatographie — HPLC,* Springer Verlag, Berlin-Göttingen-Heidelberg-New York 1977.
[403] J. J. Kirkland: *Modern Practise of Liquid Chromatography,* Wiley-Interscience, New York 1971.
[404] H. M. McNair, *Int. Lab.* **1980,** no. 5/6, 51–59.
[405] L. P. Turner et al., *J. Am. Oil Chem. Soc.* **53** (1976) 691–694.
[406] H. Henke, *Tenside Deterg.* **15** (1978) 193–195.
[407] R. Murphy, A. C. Selden, M. Fisher, E. A. Fagan, V. S. Chadwick, *J. Chromatogr.* **211** (1981) 160–165.
[408] J. H. van Dijk et al., *Tenside Deterg.* **12** (1975) 261–263.
[409] D. Kirkpatrick, *J. Chromatogr.* **121** (1976) 153–154.
[410] Th. Wolf, D. Semionow, *J. Soc. Cosmet. Chem.* **24** (1973) 363–370.
[411] A. Nozawa, T. Ohnuma, *J. Chromatogr.* **187** (1980) 261–263.
[412] K. Nakamura, Y. Morikawa, I. Matsumoto, *J. Am. Oil Chem. Soc.* **58** (1981) Jan., 72–77.
[413] H. Ullner, I. König, C. Sander, Schwenk, *Tenside Deterg.* **17** (1980) 169–170.
[414] H.-U. Ehmcke, H. Kelker, K.-H. König, H. Ullner, *Fresenius' Z. Anal. Chem.* **294** (1979) 251–261.
[415] D. R. Zornes et al., *Petr. Eng. J.* **18** (1978) 207–218.
[416] W. Winkle, M. Köhler, *Chromatographia* **13** (1980) 357–363.
[417] M. Pestemer: *Anleitung zum Messen von Absorptionsspektren im Ultraviolett und Sichtbaren,* Thieme Verlag, Stuttgart 1964.
[418] E. J. Stearns: *The Practice of Absorption Spectrophotometry,* Wiley-Interscience, New York 1969.
[419] C. N. R. Rao: *Ultraviolet and Visible Spectroscopy,* 3rd ed., Butterworths, London 1975.
[420] G. Milazzo et al., *Anal. Chem.* **49** (1977) no. 6, 711–717.
[421] W. J. Weber, J. C. Morris, W. Stumm, *Anal. Chem.* **34** (1962) 1844–1845.
[422] R. M. Kelley, E. W. Blank, W. E. Thompson, R. Fine, *ASTM Bull.* **TP 90** (1959) 70–73.

[423] M. Uchijama, *Water Res.* **11** (1977) 205–207.
[424] E. Heinerth, H. G. van Raay, G. Schwarz, *Fette Seifen Anstrichm.* **62** (1960) 825–826.
[425] V. W. Reid, T. Alston, B. W. Young, *Analyst (London)* **80** (1955) 682–689.
[426] F. N. Stewart et al., *Anal. Chem.* **31** (1959) no. 11, 1806–1808.
[427] H. J. Hediger: *Methoden der Analyse in der Chemie,* vol. 11: „Infrarotspektroskopie," Akademische Verlagsges., Frankfurt 1971.
[428] L. J. Bellamy: *Ultrarot-Spektrum und chemische Konstitution,* D. Steinkopff Verlag, Darmstadt 1966.
[429] Weissberger: *Technique of Organic Chemistry,* vol. IX: "Chemical Application of Spectroscopy," Interscience Publ. Inc., New York 1956.
[430] H. König: *Neuere Methoden zur Analyse von Tensiden,* Springer Verlag, Berlin-Göttingen-Heidelberg-New York 1971.
[431] E. Kunkel, *Mikrochim. Acta* **1977,** 227–240.
[432] H. Günzler, H. Böck: *IR-Spektroskopie,* Verlag Chemie, Weinheim 1975.
[433] G. Socrates: *Infrared Characteristic Group Frequencies,* J. Wiley & Sons, New York 1980.
[434] H. G. van Raay, M. Teupel, *Fette Seifen Anstrichm.* **75** (1973) 572–578.
[435] C. D. Frazee, R. O. Crister, *J. Am. Oil Chem. Soc.* **41** (1964) 334–335.
[436] H. Hellmann, *Fresenius' Z. Anal. Chem.* **294** (1979) 379–384.
[437] H. Hellmann, *Fresenius' Z. Anal Chem.* **293** (1978) 359–363.
[438] P. Friese, *Fresenius' Z. Anal Chem.* **305** (1981) 337–346.
[439] H. Günther: *NMR-Spektroskopie – Eine Einführung,* Thieme Verlag, Stuttgart 1973.
[440] H. König, *Fresenius' Z. Anal. Chem.* **251** (1970) 225–262.
[441] A. Mathias, N. Mellor, *Anal. Chem.* **38** (1966) 472–477.
[442] M. M. Crutchfield, R. R. Irani, J. T. Yoder, *J. Am. Oil Chem. Soc.* **41** (1964) 129–132.
[443] H. Budzikiewicz: *Massenspektrometrie – Eine Einführung,* Verlag Chemie, Weinheim 1972 (Taschentext 5).
[444] M. Köhler, M. Höhn, *Chromatographia* **9** (1976) 611–617.
[445] H. A. Boekenoogen: *Oils, Fats and Fat Products,* vol. 2, Interscience Publ., New York 1968.
[446] D. Jahr, P. Binnemann, *Fresenius' Z. Anal. Chem.* **298** (1979) 337–348.
[447] W. McFadden: *Techniques of Combined Gas Chromatography/Mass Spectrometry,* J. Wiley & Sons, London 1973.
[448] R. ter Heide, N. Provatoroff, P. C. Traas, P. J. Valois, *J. Agric. Food Chem.* **23** (1975) 950–957.
[449] M. Köhler, *Chromatographia* **8** (1975) 685–689.
[450] F. Ehrenberger, S. Gorbach: *Methoden der organischen Elementar- und Spurenanalyse,* Verlag Chemie, Weinheim 1973.
[451] J. Monar, *Mikrochim. Acta* **2** (1965) 208–250.
[452] R. Wickbold, *Angew. Chem.* **69** (1957) 530–533.
[453] F. Ehrenberger, *GIT Fachz. Lab.* **21** (1977) 944.
[454] R. Kaiser: *Quantitative Bestimmung Organischer Funktioneller Gruppen,* Akademische Verlagsges., Frankfurt 1966.
[455] *Technicon-Symposia, Technicon Intern. Congress 1969,* vol. II, White Plains, Mediad 1970.
[456] D. P. Lundgren, *Anal. Chem.* **32** (1960) 824–828.
[457] *Technicon Industrial Method 3–68 W,* Technicon Corporation Tarrytown/New York 10591.
[458] W. Gohla et al., *GIT Fachz. Lab.* **23** (1979) 89–96.
[459] H. Pfeiffer, K. H. Lange, *Fette Seifen Anstrichm.* **75** (1973) 438–442.
[460] E. Heinerth, *Tenside Deterg.* **7** (1970) 23–24.
[461] H. Grossmann, *Seifen Öle Fette Wachse* **101** (1975) 521–524.
[461 a] *Fette Seifen Anstrichm.* **77** (1975) 80.
[462] L. Noll, H. Stache, *Tenside Deterg.* **17** (1980) 1.

[463] "CESIO-Generalversammlung 1980," *Tenside Deterg.* **17** (1980) 201–294.
[464] Manufacturer: Atlas Electric Devices Comp., 4114 N. Ravenswood Ave., Chicago.
[465] Manufacturer: Heraeus-Original Hanau, D-6450 Hanau.
[466] Manufacturer: United States Testing Co. Inc., Hoboken N.J.
[467] DIN 53 902, Part 1: *Bestimmung des Schäumvermögens, Lochscheibenschlagverfahren.*
[468] DIN 53 902, Part 2: *Bestimmung des Schäumvermögens, modifiziertes Ross-Miles-Verfahren.*
[469] H. Machemer, W. Griess, H. Mugele, *Fette Seifen* **54** (1952) 769–780.
[470] Manufacturer: Wäschereiforschung Krefeld, WFK-Testgewebe GmbH, Adlerstraße 44, D-4150 Krefeld; EMPA Eidgenössische Materialprüfungs- und Versuchsansalt, Unterstraße 11, CH-9001 St. Gallen; Testfabrics Inc., 200 Blackford Ave., Middlesex, N.J.
[471] H. Stache in: *Tensid-Taschenbuch,* 2nd ed., Hanser Verlag, München-Wien 1981, pp. 460–461.
[472] H. Stache in: *Tensid-Taschenbuch,* 2nd ed., Hanser Verlag, München-Wien 1981, pp. 462–466.
[473] U. Sommer, H. Milster, *Seifen Öle Fette Wachse* **103** (1977) 295.
[474] H. Harder, D. Arends, W. Pochandke, *Seifen Öle Fette Wachse* **102** (1976) 421–426.
[475] H. Krüssmann, *J. Am. Oil Chem. Soc.* **55** (1978) 165.
[476] H. Harder, *Seifen Öle Fette Wachse* **94** (1968) 789–794, 825.
[477] H. Harder, D. Arends, W. Pochandke, *Seifen Öle Fette Wachse* **103** (1977) 180–183.
[478] DIN 53 990: *Waschmittel zum Waschen von Textilien, Empfehlungen für die vergleichende Prüfung von Gebrauchseigenschaften.*
[479] ISO TC 91-4319-1977: *Surface active agents – Detergents for washing fabrics – Guide for comparative testing of performance.*
[480] *J. Am. Oil Chem. Soc.* **61** (1984) 8.
[481] R. von der Grün, S. Scholz-Weigl in: *Tensid-Taschenbuch,* 2nd ed., Hanser Verlag, München-Wien 1981, pp. 506–521.
[482] G. Jakobi, P. Krings, E. Schmadel in J. Falbe: *Surfactants in Consumer Products,* Springer Verlag, Heidelberg 1987, pp. 504–512.
[483] B. Werdelmann in: *Proceedings of the World Surfactants Congress München,* vol. I, Kürle Verlag, Gelnhausen 1984, p. 3.
[484] Hauptausschuß Phosphate: *Phosphor, Wege und Verbleib in der Bundesrepublik Deutschland.* Verlag Chemie, Weinheim-New York 1978, pp. 29–30.
[485] P. Berth, *Tenside Deterg.* **19** (1982) 100–103.
[486] P. Berth, P. Krings, H. Verbeek, *Tenside Deterg.* **22** (1985) 169.
[487] Henkel KGaA, Volkswirtschaftliche Abteilung, unpublished data.
[488] A. L. de Jong, *Soap Cosmet. Chem. Spec.* **1983,** April, 31–40.
[489] R. Schulze-Rettmer, *Reiniger Wäscher* **27** (1974) no. 7, 19–20.
[490] R. Schulze-Rettmer, *Seifen Öle Fette Wachse* **102** (1976) 427–430.
[491] R. Wagner, *GWF Gas Wasserfach; Wasser/Abwasser* **119** (1978) 235–242.
[492] H. Kemmerling, D. Dames, *Reiniger Wäscher* **33** (1980) no. 3, 21–28.
[493] R. Schulze-Rettmer, *Reiniger Wäscher* **28** (1975) no. 4, 24–26.
[494] R. Schulze-Rettmer, *Reiniger Wäscher* **24** (1971) no. 6, 42–43.
[495] R. Schulze-Rettmer in *GDCh-Fortbildungskurs Roetgen,* pp. 174–184 (1980).
[496] H. Krüssmann, H. G. Hloch, *Seifen Öle Fette Wachse* **107** (1981) 436–442.
[497] R. Schulze-Rettmer, *Wäscherei Reinigungsprax.* **26** (1977) 6–10.
[498] Europäisches Übereinkommen über die Beschränkung der Verwendung bestimmter Detergentien in Wasch- und Reinigungsmitteln vom 16.9.1968 (BGBl. 1972 I, p. 1718).
[499] Richtlinie des Rates vom 22.11.1973 zur Angleichung der Rechtsvorschriften der Mitgliedstaaten über Detergentien, 73/404/EWG, EG-ABl. L 347, p. 51; Council Directive 73/404/EEC, OJ L 347, p. 51.
[500] Richtlinie des Rates vom 22.11.1973 zur Angleichung der Rechtsvorschriften der Mitgliedstaaten über die Methoden zur Kontrolle der biologischen Abbaubarkeit anionischer

grenzflächenaktiver Substanzen, 73/405/EWG, EG-ABl. L 347, p. 33; Council Directive 73/405/EEC, OJ L 347, p. 33.
[501] J. Au, *Tenside Deterg.* **18** (1981) 280–285.
[502] Gesetz über die Umweltverträglichkeit von Wasch- und Reinigungsmitteln (Waschmittelgesetz) vom 20.8.1975 (BGBl. I, p. 2255).
[503] W. Langer, *Tenside Deterg.* **18** (1981) 285–286.
[504] Verordnung über die Abbaubarkeit anionischer und nichtionischer grenzflächenaktiver Stoffe in Wasch- und Reinigungsmitteln vom 30.1.1977 (BGBl. I, p. 244) (geändert durch Verordnung vom 18.6.1980, BGBl. I, p. 706).
[505] Verordnung zur Änderung der Verordnung über die Abbaubarkeit anionischer und nichtionischer grenzflächenaktiver Stoffe in Wasch- und Reinigungsmitteln vom 18.6.1980 (BGBl. I, p. 706).
[506] Richtlinie des Rates vom 31.3.1982 zur Änderung der Richtlinie 73/405/EWG zur Angleichung der Rechtsvorschriften der Mitgliedstaaten über die Methoden zur Kontrolle der biologischen Abbaubarkeit anionischer grenzflächenaktiver Substanzen, 82/243/EWG, EG-ABl. L 109/18, 22.4.1982; Council Directive 82/243/EEC, OJ L 109, p. 18.
[507] W. K. Fischer, K. Winkler, *Vom Wasser* **47** (1976) 81–129.
[508] H. Hellmann, *Tenside Deterg.* **15** (1978) 291–294.
[509] H. Hellmann, *GWF Gas Wasserfach: Wasser/Abwasser* **122** (1981) 158–162.
[510] W. K. Fischer, *Tenside Deterg.* **17** (1980) 250–261.
[510a] P. Berth, P. Gerike, P. Gode, B. Steber in: *Proceedings of the World Surfactants Congress München,* vol. I, Kürle Verlag, Gelnhausen 1984, p. 227.
[511] Hauptausschuß Phosphate: *Phosphor, Wege und Verbleib in der Bundesrepublik Deutschland,* Verlag Chemie, Weinheim-New York 1978.
[512] J. Kandler, G. Sorbe, *Seifen Öle Fette Wachse* **105** (1979) 51–54.
[513] Verordnung über Höchstmengen für Phosphate in Wasch- und Reinigungsmitteln (Phosphathöchstmengenverordnung — PHöchstMengV) vom 4.6.1981 (BGBl. I, p. 664–665).
[514] W. K. Fischer, P. Gerike, R. Schmid, *Zeitschr. f. Wasser Abwasser Forsch.* **7** (1974) 99–118.
[515] P. Berth, W. K. Fischer, R. Schmid, *Tenside Deterg.* **14** (1977) 51–53.
[516] H. C. Kemper, R. J. Martens, J. R. Nool, C. E. Stubbs, *Tenside Deterg.* **12** (1975) 47.
[517] P. Berth, *J. Am. Oil Chem. Soc.* **55** (1978) 52–57.
[518] P. Berth, M. J. Schwuger, *Tenside Deterg.* **16** (1979) 1–12.
[519] P. Berth, W. K. Fischer, Ch. Gloxhuber, K. Hachmann, A. Roland, R. Schmid, M. J. Schwuger, *World Conf. on Soaps and Detergents, Montreux 1977, special issue of J. Am. Oil Chem. Soc.,* 1978.
[520] Umweltbundesamt: *Die Prüfung des Umweltverhaltens von Natrium-Aluminium-Silikat Zeolith A als Phosphatersatzstoff in Wasch- und Reinigungsmitteln.* Materialien 4/1979, E. Schmidt-Verlag, Berlin.
[521] W. A. Roland, *Forum Städte Hyg.* **30** (1979) 131–137.
[522] W. A. Roland, *Umwelt* **10** (1980) 237–242.
[523] P. Koppe, *Z. Wasser Abwasser Forsch.* **9** (1976) 153–161.
[524] K. E. Quentin, *Tenside Deterg.* **18** (1981) 269–274.
[525] „Zur Frage der Abbaubarkeit schaumarmer, nicht-ionischer Tenside in Reinigungsmitteln für Gewerbe und Industrie", „Kationische Tenside – Umweltaspekte", *Tenside Deterg.* **19** (1982) 121–196.
[526] W. Janicke, G. Hilge, *Tenside Deterg.* **16** (1979) 117–122.
[527] P. Gerike, *Tenside Deterg.* **19** (1982) 162–163.
[528] P. Gerike, W. K. Fischer, W. Jasiak, *Water Res.* **12** (1978) 1117–1122.
[529] L. Huber, *Tenside Deterg.* **19** (1982) 178–180.
[530] W. Kupfer, *Tenside Deterg.* **19** (1982) 158–161.

[531] J. Waters, W. Kupfer, *Anal. Chim. Acta* **85** (1976) 241–251.
[532] J. Waters, *Tenside Deterg.* **19** (1982) 177.
[533] P. Gerike, W. K. Fischer, W. Holtmann, *Tenside Deterg.* **13** (1976) 249.
[534] F. Dietz, *GWF Gas Wasserfach: Wasser/Abwasser* **116** (1975) 301–308.
[535] G. Graffmann, P. Kuzel, H. Nösler, G. Nonnenmacher, *Chem. Ztg.* **98** (1974) 499–504.
[536] Acad. Eng. Science: *Boron from the Standpoint of Environment,* IVA-Report 33, Stockholm 1970.
[537] G. Müller, U. Nagel, I. Purba, *Chem. Ztg.* **102** (1978) 169–178.
[538] H. Hellmann: *Gewässerkundliche Untersuchungen über die Dynamik des Umsatzes von Phosphat, Nitrat und Borat im Rhein,* Forschungsvorhaben Nr. 8/74/20 Schlußbericht vom 21.3.1977, Bundesanstalt für Gewässerkunde, Koblenz.
[539] W. K. Fischer, *Tenside Deterg.* **12** (1975) 53–64.
[540] R. Anliker, G. Müller: *Fluorescent Whitening Agents,* Thieme Verlag, Stuttgart 1975.
[541] R. F. Sharp, H. O. W. Eggins, *Int. Biodeterior. Bull.* **6** (1970) 19–26.
[542] Richtlinie des Rates vom 18.9.1979 zur sechsten Änderung der Richtlinie 67/548 EWG zur Angleichung der Rechts- und Verwaltungsvorschriften für die Einstufung, Verpackung und Kennzeichnung gefährlicher Stoffe, 79/831/EWG, EG-ABl. L 259, p. 10; Council Directive 79/831/EEC, OJ L 259, p. 10.
[543] Gesetz zum Schutz vor gefährlichen Stoffen (Chemikaliengesetz – ChemG) vom 16.9.1980 (BGBl. I, p. 1718).
[544] M. J. Schwuger, F. G. Bartnik in Ch. Gloxhuber (ed.): *Anionic Surfactants, Biochemistry, Toxicology, Dermatology,* Surfactant Science Ser., vol. 10, Marcel Dekker, New York 1981, pp. 1–49.
[545] J. G. Black, D. Howes in Ch. Gloxhuber (ed.): *Anionic Surfactants, Biochemistry, Toxicology, Dermatology,* Surfactant Science Ser., vol. 10, Marcel Dekker, New York 1981, pp. 51–85.
[546] R. B. Drotman in V. A. Drill, P. Lazar (eds.): *Cutaneous Toxicity,* Academic Press, New York, 1977, pp. 96–109.
[547] R. B. Drotman, *Toxicol. Appl. Pharmacol.* **52** (1980) 38.
[548] B. Isomaa, *Food Cosmet. Toxicol.* **13** (1975) 231.
[549] A. Siwak, M. Goyer, J. Perwak, P. Thayer in K. L. Mittal, E. J. Fendier (eds.): *Solution Behavior of Surfactants,* vol. I, Plenum Publ. Corp., New York 1982, p. 161.
[550] R. A. Cutler, H. P. Drobeck in E. Jungermann (ed.): *Cationic Surfactants,* Surfactant Science Ser. vol. 5, Marcel Dekker, New York 1970, p. 527.
[551] D. L. Opdyke, C. M. Burnett, *Proc. Sci. Sect. Toilet. Goods Assoc.* **44** (1965) 3.
[552] W. Kästner in Ch. Gloxhuber (ed.): *Anionic Surfactants, Biochemistry, Toxicology, Dermatology,* Surfactant Science Ser., vol. 10, Marcel Dekker, New York 1981, pp. 139–307.
[553] Ch. Gloxhuber, M. Potokar, S. Braig, H. G. van Raay et al., *Fette Seifen Anstrichm.* **76** (1974) 126.
[554] R. D. Swisher, *Arch. Environ. Health* **17** (1968) 232.
[555] W. B. Coate, W. M. Busey, W. H. Schoenfisch, N. M. Brown et al., *Toxicol. Appl. Pharmacol.* **45** (1978) 477.
[556] W. Kissler, K. Morgenroth, W. Weller, *Prog. Respi. Res.* **15** (1981) 121.
[557] M. S. Potokar in Ch. Gloxhuber (ed.): *Anionic Surfactants, Biochemistry, Toxicology, Dermatology,* Surfactant Science Ser., vol. 10, Marcel Dekker, New York 1981, pp. 87–126.
[558] K. Oba in Ch. Gloxhuber (ed.): *Anionic Surfactants, Biochemistry, Toxicology, Dermatology,* Surfactant Science Ser., vol. 10, Marcel Dekker, New York 1981, pp. 327–403.
[559] Y. Yam, K. A. Booman, W. Broddle, L. Geiger et al., *Food Chem. Toxicol.* **22** (1984) no. 9, 761–769.
[560] Ch. Gloxhuber, *Fette Seifen Anstrichm.* **74** (1972) 49.

[561] Ch. Gloxhuber, M. Potokar, W. Pittermann, S. Wallat et al., *Food Chem. Toxicol.* **21** (1983) no. 2, 209–220.
[562] G. A. Nixon, *Toxicol. Appl. Pharmacol.* **18** (1971) 398–406.
[563] W. R. Michael, J. M. Wakim, *Toxicol. Appl. Pharmacol.* **18** (1971) no. 2, 407–416.
[564] J. A. Budny, J. D. Arnold, *Toxicol. Appl. Pharmacol.* **25** (1973) no. 1, 43–53.
[565] G. A. Nixon, E. V. Buehler, R. J. Nieuwenhuis, *Toxicol. Appl. Pharmacol.* **21** (1972) no. 2, 244–252.
[566] National Cancer Institute, NCI Techn. Rep. Ser. no. 6, DHEW Publication no. (NHI) 77–806 (1977).
[567] P. S. Thayer, C. J. Kensler, A. D. Little, *CRC Crit. Rev. Environ. Control* **3** (1973) 335–340.
[568] Ch. Gloxhuber, *Med. Welt* **19** (1968) 351–357.
[569] Ch. Gloxhuber, J. Malaszkiewicz, M. Potokar, *Fette Seifen Anstrichm.* **73** (1971) 182–189.
[570] G. J. Schmitt, *Z. Hautkr.* **49** (1974) 901.
[571] F. Coulston, F. Korte (eds.): *Fluorescent Whitening Agents,* Georg Thieme Verlag, Stuttgart 1975.
[572] A. W. Burg, M. W. Rohovsky, C. J. Kensler, *CRC Crit. Rev. Environ. Control* **7** (1977) 91–120.
[573] CIRFS (Comité International de la Rayon et des Fibres Synthétiques, Paris): *Data on Synthetic Fibers,* 1984.
[574] R. Weber, *Seifen Öle Fette Wachse* **95** (1969) 885–891.
[575] R. Weber, *Textilveredlung* **15** (1980) 380–385.
[576] W. Graupner, F. Moczarski, H. Puderbach, *Melliand Textilber. Int.* **55** (1974) 1039–1041.
[577] G. Wildbrett: *Technologie der Reinigung im Haushalt,* Verlag E. Ulmer, Stuttgart 1981, pp. 145–150.
[578] *Laundry Guide for Whirlpool Automatic Washers* Pert. No. 358 770 Rev. B.
[579] H. Schmidt: Forschungsbericht des Landes Nordrhein-Westfalen no. 892, Westdeutscher Verlag, Köln-Opladen 1960.
[580] H. Pichert, *Seifen Öle Fette Wachse* **104** (1978) 509–512.
[581] R. Puchta, W. Grünewälder: *Textilpflege, Waschen und Chemischreinigen,* Verlag Schiele und Schön, Berlin 1973, pp. 35–44.
[582] H. Milster, U. Sommer, *Materialprüfung* **21** (1979) 232–236.
[583] Zentralverband der Elektrotechnischen Industrie e.V. (ZVEI): *Zahlenspiegel der deutschen Hausgeräteindustrie, 1984.*
[584] Hauptberatungsstelle für Elektrizitätsanwendung e.V. (HEA): Messe-Schnellinformation, Domotechnika 1985.
[585] J. Wangler, *Appliance* **42** (1985) Sept., 29–33.
[586] *Appliance* **42** (1985) April, 41.
[587] Blick durch die Wirtschaft, Frankfurter Allgemeine Zeitung, 22.7.1983.
[588] Information of the producer of washing machines and tumbler dryers, Miele, West Germany 1984.
[589] H. Strone, W. Löbrich, R. Senf, K. D. Wetzler: *Bildungsfibel,* Part 1. Deutscher Textilreinigungsverband, Bonn 1980, pp. 133–149.
[590] Henkel KGaA: *Industrielles Waschen,* Rheinisch-Bergische Druckerei GmbH, Düsseldorf 1978, pp. 24–35.

Index